The End
of Adam and Eve

The End of Adam and Eve

Theology and the Science of Human Origins

====================

Ron Cole-Turner

TheologyPlus Publishing

TheologyPlus Publishing
616 N Highland Ave., Pittsburgh PA 15206

Copyright 2016 Ron Cole-Turner
All rights reserved. No part of this book may be reproduced by an means without the written permission of the copyright holder.

ISBN 978-0-9980686-0-2 (print)
ISBN 978-0-9980686-1-9 (Kindle edition)
ISBN 978-1-3706235-6-3 (other formats)

Cover Photo: Faithful photographic image of Hendrick Goltzius (1558-1617), "The Fall of Man," 1616, now at the National Gallery of Art, Washington. Oil on canvas. Photo by Daderot. In the public domain.

Scripture quotations come from the Revised Standard Version of the Bible, copyright 1952 [2nd edition, 1971] by the Division of Christian Education of the National Council of the Churches of Christ in the United States of America. Used by permission. All rights reserved.

Digital versions of this book, including PDF, are available free.

Readers are encouraged to visit www.theologyplus.org for updates about the scientific research presented in this book or to comment about its contents.

Acknowledgements

This book began as a few rough notes for students daring enough to take my course on theology and evolution. I am grateful to them for their tough questions and their sustained interest. My appreciation also extends to my colleagues at Pittsburgh Theological Seminary, many of whom surely wondered what kind of theologian comes running out of his office announcing the latest discoveries about Neandertals.

I am also grateful to Craig Story, biologist at Gordon College, who read an early version of the text and offered many helpful comments. Bill Brown and Martha Moore-Keish not only invited me to speak to students at Columbia Theological Seminary but used an early draft of the book in one of their classes. I am grateful to them and to the students whose comments helped sharpen its argument.

Above all I am indebted to my wife, Rebecca, for encouraging me in this project. Not only did she help in cleaning up the manuscript, but her advice was key to getting the title right.

Contents

1. Why This Book Is Needed ... 11
   Adam and Eve and Darwin: The Challenge to Christian Faith Today 13
   What Science Can and Cannot Tell Us 16
   My Two Goals 18

2. Our First Three Million Years ... 21
   Bones of Contention 22
   Our First Leading Lady 23
   If Teeth Could Talk 25
   Ardi and the Last Common Ancestor 28
   Before Ardi 32
   Evolving Debates 35

3. Lucy and the Next Two Million Years of the Human Story ... 39
   Lucy and Her Family 40
   Next of Kin 43
   Toes and Toddlers 44
   The Last Pre-Human 47
   Searching for Surprises 50

4. The Dawn of Technology ... 53
   Cores and Flakes 54
   Rolling Stones 58
   Fire 60
   Four Unanswered Questions 63

5. Genesis and Exodus: The Genus *Homo* Travels the World ... 67
   The Human Question Mark 69
   The First Humans 72
   Walking Tall, But Which Way? 77
   What Were They Like? 79
   The Importance of Being Adaptable 81
   After Homo Erectus: *Bigger Brains and Diverging Forms* 84

6. Are We Modern Yet?   87
   Anatomically Modern   88
   Behaviorally Modern   93
   Humanly Modern   98
   From Tools to Art   100
   Cave Paintings and Musical Instruments   103

7. Born Again Neandertals   107
   Who Were the Neandertals?   108
   Neandertal Culture   111
   Disappearing or Assimilating?   115
   Getting Bones to Talk   118
   The Legacy of Interbreeding   120
   Human Emergence through Interbreeding Lineages   122

8. When Adam and Eve Disappear   127
   A New Story, or More of the Same?   128
   Why We Need to Learn from the Past   131
   Racism and Human Unity   133
   Christianity and Polygenism   136
   Human Souls   139
   Are Humans Fallen?   143
   Options for Christian Theology   148

9. Evolving in the Image of God   151
   Challenges Old and New   153
   Defining the Image   154
   Refining the Image   160
   The Image of God and Human Uniqueness   163
   How Does a Spiritual Animal Evolve?   166
   New Paths, New Meanings   170

10. Christ Makes Us All One   173
    Christ Completes Humanity   175
    Christ Unifies Humanity   179
    The Convergent Christ   181
    Christ Defines…and Undefines   185
    Christ of the Future   187

References   195

1

Why This Book Is Needed

Like most American kids growing up in the 1950s, I believed I was a direct descendant of Adam and Eve. That is what my parents believed, what everyone around me believed, and what the Bible said. I even had a picture of Adam and Eve in my Bible storybook. They were hiding discretely behind a bush, but I could see that they had light skin and brown hair just like me. There was no reason to think anything else, and it was all very reassuring.

One day in fifth grade, we opened our social studies book to the first chapter and read aloud about early human beings. The book contained nothing about evolution, but this tidbit of scientific secularism caught my eye. The textbook said that in the beginning, human beings spoke many languages. I was horrified.

I happened to be in a church-based school. Even so, false teaching can enter in and must be exposed wherever it is found, and I was ready to meet the challenge.

I raised my hand. "Not true," I said, pointing to the Bible. "Just look at Genesis 11:1," I said. "It says right here that the whole earth spoke only one language until God confused them because they built the tower of Babel in order to get to heaven on their own." The teacher looked puzzled, but she had to agree with me even though I think she knew better. After all, she was teaching in a Christian school and probably needed the job.

In a way, I am writing this book for her and for all the other decent, intelligent, sincere Christians who do not want to be forced to choose between their faith and their minds. This is my way of making up for some of the abuse they get from people like I used to be, people who think that the Bible is the sole authority on questions like the origin of us humans and our languages.

Today, I find the scientific account of human origins to be truly exciting, riveting, and ennobling. Following the science makes me a better Christian. It makes me more grateful to the God who not only creates us human beings but comes into our humanity in Jesus Christ. The truth of the incarnation is all the more glorious in light of the truth of evolution.

Never has this been truer than today. The field of human origins research has expanded dramatically in just the past few years. Many of the discoveries discussed in this book are dated 2008 or later. These include perhaps the most dramatic finding of all. Now we know that we are descended partly from Neandertals and probably from many other forms of extinct ancestors. As the story of our past is filled in with more detail, the surprises and the new questions grow exponentially. I find this personally intriguing, and I have come to see myself as a living bearer of this complex ancestry, a descendant of earlier forms of humanity that lived millions of years ago. It is as if their lives still echo in my DNA and in the genetic underpinnings of my anatomy and my behavior. In a way, they are not dead at all. They live on in us today.

All the way through college, I rejected evolution. I would have found it abhorrent to think that I was descended from "cave men" and other earlier forms of humanity. It took years to change my mind. In fact, I don't remember actually changing it. But little by little as time passed, my ideas changed and I began to see "Adam and Eve" in a new way. Now I have reached the point that, instead of needing to protect the Bible from science, I want to see how far I can go in making sense of the Christian faith in light of science.

I want my faith to be alive, spiritually and intellectually. This means having the courage to rethink completely what traditional Christian faith says about humanity. For some, this is too bold, too audacious.

But is there really any other choice? *I want to think like a Christian who lives in this world today, knowing what we know through the sciences and trying honestly to put it all together.*

Adam and Eve and Darwin: The Challenge to Christian Faith Today

Why is that important? Why is it important to me (and I hope to you) to try to think through our faith in light of the natural sciences? Why is it necessary to rethink our theology of human origins in response to science?

I can think of two main reasons. First, when it comes to knowing anything about nature, *science is an essential key to our understanding.* I cannot fully understand humans without science. Second, when it comes to understanding humanity as a part of my Christian outlook, *paying attention to evolution enriches Christian faith.* Let's think for a minute about these two reasons.

First, *science is an essential key to understanding nature.* More than ever, the many disciplines and tools that make up the scientific enterprise today are yielding new insight into every dimension of nature. And since we human beings are part of nature, science reveals us to ourselves. More profoundly and more quickly than ever before, science is uncovering the secrets of our biology. This includes the details of human brains and bodies, the intricacies of what goes on inside our cells, and the ways in which our genes help define us. Quite simply, if I want to know about myself as a living creature, I must include the insights of science if I want the clearest, most detailed, and most reliable view.

In practice, most of us agree with that. When a loved one is seriously ill, we want the best medicine based on the latest research and making use of the most advanced technology.

But something odd happens when it comes to evolution and human origins. A surprisingly large number of people who respect science in general seem to reject it when it touches on questions about where we came from. They reject the idea that human beings evolved. Nearly half of all Americans say they believe that human beings were directly created by God sometime in the last 10,000 years, according to a Gallup survey reported in 2012.

But in the same year, the Barna Group looked into reasons why young adults are leaving the church. The report, aptly titled *You Lost Me*, identifies a perceived conflict between faith and science as a key reason for dropping out. "One of the reasons young adults feel disconnected from church or from faith is the tension they feel between Christianity and science. The most common of the perceptions in this arena is 'Christians are too confident they know all the answers' (35%)."

The findings continue with this sobering note: "Three out of ten young adults with a Christian background feel that 'churches are out of step with the scientific world we live in' (29%). Another one-quarter embrace the perception that 'Christianity is anti-science' (25%). And nearly the same proportion (23%) said they have 'been turned off by the creation-versus-evolution debate'" (Kinnaman and Hawkins 2011).

Taken together, these two surveys are disturbing. On the one hand, nearly one in two Americans rejects the idea that humans have evolved. On the other hand, young people are leaving church by the millions because they think that churches are anti-evolution. The fact is that for the most part, the churches themselves are not anti-evolution. And my hunch is that many people who say they reject human evolution either have distorted ideas about what it means, or they think that evolution somehow undermines their Christian faith.

I disagree. Evolution does not undermine authentic Christian faith. On the contrary, evolution enriches our faith. And that takes me to my second reason why I am taking on this challenge: *paying attention to evolution enriches Christian faith*.

Not everyone feels that way. Many Christians sincerely think that evolution threatens their faith. I know because I used to be one of them. But now I am convinced that God wants us to know the truth about creation and especially about ourselves. In this book, however, I will not try to argue that Christians should take evolution seriously. Others have done this already. For example, Ted Peters and Marty Hewlett have provided an excellent and readable introduction to the broad questions about the relationship between evolution and Christianity, showing that the whole idea of "war" between them is not necessary. In *Evolution from Creation to New Creation*, Peters and Hewlett offer an insightful and sympathetic interpretation of creationism and intelligent design, and they argue for an alternative that retains and even highlights theology's central core while reconciling faith with science (Peters and Hewlett 2003). Another good book is Darrel Falk's *Coming to Peace with Science* (Falk 2004). I recommend these books to anyone who is struggling over the question of whether a Christian can accept the findings of evolutionary biology.

Here in this book, my focus is not on evolution in general but on human origins in particular. I am convinced that God wants us to have the best possible insight into what it means to be human. In fact, I have a hunch that God wants us to be excited by the process of discovery. God made us curious, eager to learn, always open to new ideas.

This is especially important today, when the pace of discovery is so fast it sometimes makes our heads spin. New research is constantly expanding our knowledge of the natural world, sparking new ideas, challenging old ideas, and renewing our thinking about big questions, like *what does it mean to be human?* or *why are we here?* For Christians, these are theological questions. To answer them fully, we have to refer to God. But science also has insight to offer.

I believe that our best answers to questions like these are the answers that are rooted in a rich theology and informed by the most recent scientific insight. Science is a dynamic and exciting field, constantly generating new perspectives. As much as I love old texts, and as much as I am struck by the profundity of the abiding message of the Bible, I find myself prodded intellectually by new reports in scientific journals. It is mentally refreshing to keep on learning, and doing so is even more exciting for me when I am part of a global community of scientists and scholars who are all discovering roughly the same things at roughly the same time.

All this feeds my sense of calling as a Christian. It is just too easy to undersell God. The living God is always bigger, more imaginative, and more surprising that we think. I really believe that many people today are turned off to Christianity because too many Christians have a small, tired, somewhat pathetic view of God. They feel they have to protect their God from things like science and evolution, as if God doesn't know how we got here! I don't think it is possible to hold too big a view of God.

But this is a risk not everyone will take. I want to be as honest about this as I can. When Christian theology takes science on board, a lot changes. After all, Christian theology has been around for 2000 years. Only in the past 150 years has it even been suggested that humans *evolved*. And only in the past few years have we learned most of the details of what this means. So all that wonderful theology of the past was based on what I see as outmoded views of humanity.

This is a serious challenge because for Christianity, humanity is a pretty big deal. Not only do we believe that human beings are created in the image of God. We also believe that in Jesus Christ, God enters into our world to live among us in a tangibly present way, in human form. For a serious-minded Christian, there is no avoiding the question: what do we mean by "human"?

What Science Can and Cannot Tell Us

Now of course, as interesting as the peer-reviewed reports of scientists might be, they are not *THE TRUTH*. They are provisional findings that are rigorously obtained by painstaking observations of nature. Sometimes, the best scientific findings raise more questions than they answer. Together the findings that come from scientific research provide a composite picture, a set of insights that must be taken broadly into account when we try to put together any kind of comprehensive view of reality, including a theological view.

Science always faces limits. For example, everyone studying human origins wishes we had more bones, more fossils, more sites, and more ancient tools and artifacts. We only have what nature has preserved and what we have had the good fortune to find. If we could find more, we would know a lot more.

Advances in technology help. One reason why the field of human origins is so exciting and dynamic at the moment is because old findings are being restudied using new tools, sometimes at the smallest scale, quite literally at the level of individual molecules. The most stunning advance in that regard has been a set of technological breakthroughs that allow researchers to reconstruct the DNA of creatures long gone.

When it comes to what happened millions of years ago, how much can we really know? The answer is always the same: not enough. It is not the quantity of facts, however, that draws us. What hooks us is that this is a profoundly human story, a story that involves all of us, a story that we are still living. In only slightly academic language, leading researchers Bernard Wood and Terry Harrison put it like this: "So why do researchers persist in trying to solve a phylogenetic problem that may well be at the limits of, or even beyond, the analytical capabilities of the data and the available methods? The reason is that our own ancestry matters to us" (Wood and Harrison 2011, 350).

At the same time, the limits of our knowledge are real. With all the advances in analysis and in spite of exciting discoveries, we still know only the tiniest fraction of what lies out there yet to be found. On the one hand, the quickening pace of discovery is electrifying the field. According to Lee Berger, "Practically never before have we seen more associated remains being discovered, in good context, so rapidly. Improved absolute dating methods and excavation techniques are allowing us to contextualize these finds" (Berger 2013b).

But what do we make of new discoveries, especially when they do not always sit well with what we think we already know? Often, new discoveries mean new arguments among experts and new debates about how to fit the fossils together into a comprehensive story. Some even wonder if they will ever sort it out. Others see each new discovery as a call for more expeditions and more analysis. Quoting Lee Berger again: "The situation we find ourselves in at present is not one of despair, but is a clarion call for more exploration and excavation, and the discovery of more and better fossils in good context" (Berger 2013b).

Never before in all of human history has our primal past been so fully revealed to us. Like it or not, we live in a time of stunning advances in the sciences of ancestral human life. The story book of human origins is just now being cracked opened and read for the first time. More discoveries, still more precise analyses, and ever greater ability to reconstruct the DNA of long-extinct forms of human life, such as the Neandertals, will pry open even more the stubborn pages of our family album.

How did we human beings come to exist, and how is it that our core features first arose? We walk on two feet. Many birds can do this, but very few mammals, and none of them do it the way we do, with speed and grace and hands left free to do other things. We have large brains that are organized to allow complex thought. Other animals have great cognitive powers, but none so complex as ours. We form reproductive pair bonds and social groups that allow us to devote extraordinary resources to the care and nurture of our young. Other animals care for their young, nursing and protecting them, but none so much as we do. How is it that we came to have these capacities and to behave in these ways? The story of human origins is the story of how these distinctive features of our humanity came into existence. Science sheds light on all these questions, and while we go on from what science reveals to ask what Christian faith also discloses, the two sources go hand in hand in showing us to ourselves.

Of all the things that make us special, our innate curiosity, our disciplined research, and our scientific prowess are among the most important. But so is our spirituality, our yearning for transcendence, our openness to enter and let ourselves be taken up into a Mystery beyond comprehension. To comprehend as far as we can, and yet at the same time always to surrender to the Incomprehensible: these are defining hallmarks of what it means to be human.

My Two Goals

Because this is true, I have two goals I hope to achieve with this book. First, I want to offer a readable, plain-English summary of recent scientific findings about human origins. There is a lot of information out there, but there is also a lot of misinformation. The trouble is that the reliable information is mostly locked up in technical reports. I mean literally locked up, in the sense that most people do not have legal access to the documents. Many journals restrict access so that most people cannot get their hands on these articles. Most people cannot see the knowledge their taxes help to create. But I also mean "locked up" in a figurative sense. Most science journal articles are written in a kind of techno-jargon code, as if to keep their secrets from mere mortals.

I have tried to make this book as accurate as possible, based on what is published in leading scientific journals through August, 2016. Wherever possible, I have drawn on the original scientific reports in the leading journals, and have tried to summarize them as accurately and as fairly as possible. Along the way, I identify my sources of information with a note in parentheses at the end of a sentence. You can find the reference to the original publication in the "References" at the end of the book.

New scientific reports are being published all the time. What I have written in this book will need to be updated before very long because of new discoveries. My plan is to provide update regularly on my website at www.theologyplus.org. I also invite readers to send suggestions for corrections. About once a year or as developments require, I plan to publish a revised edition of this book. This is my main reason for making this text available as an eBook. I want to be able to bring out new editions quickly. I also want to make the text available free to anyone who cares to read it, and the eBook format makes this possible. Finally, I want to see this book as a shared project, with readers sharing ideas and comments so that this is dynamic discussion and not a fixed text.

In addition to trying to be as accurate as possible, I have also tried to make this book readable. If you have ever read some of the journal articles, you will know that they are not always easy to read. I am hoping everyone will find my writing easy to follow. As much as possible, I try to avoid technical language. Some terms are necessary, and so I have tried to explain them as I go along.

My second reason for writing this book comes from deep within my Christian faith. As much as the world needs a *plain-English summary* of current scientific insight into the question of human origins, it is even more in need of a *plain-language Christian interpretation* of human origins in light of the latest research. More than anything, I want to try to put my scientifically-informed faith into theologically compelling terms.

Not everyone will be happy with this book. Some will say, *If you believe the gospel, then you cannot accept evolution.* And others will say: *If you accept evolution, then you cannot be religious.* I think both objections are wrong. In an odd way, they seem to agree with each other that religion and science are making the same sort of claims about facts. On that assumption, the claims made by science and faith can be mutually exclusive. If that were so, then the objectors would be right when they say that this book is a hopeless mistake.

My view is different. I see my faith as a broad, interpretative framework, a basic orientation and a grounding set of values, into which I invite all the relevant facts. As a Christian, I see human beings as created in the image of God. What makes us special is that God has come to us and joined us in our humanity in Jesus Christ. In our *evolving humanity!*

These are some of the themes that will be explored in the final three chapters of this book. Before I get started on the theology, however, I need to present the science. In the next six chapters, I want to retell the story science is revealing about how we got here.

I could start with the beginning of life on earth, over 4 billion years ago. Instead, I go back just a tiny fraction of that time. Our story begins about 7 million years ago when our ancestors split off from the ancestors of the great apes. The chimpanzees and the other great apes are our closest living relatives. We did not descend from them, nor did they descend from us. Instead, we both have evolved over the past 7 million years or so from a common ancestor, a shared ancestral group that was neither ape nor human but the ancestor of both.

This is the story of what happened next.

2

Our First Three Million Years

Today we think of Ethiopia as a land that is mostly bare and dry. Once East Africa was covered with open forests and grasslands full of wildlife. Parrots, doves, and monkeys filled the trees while antelope grazed below. If you like, you can think of it as a kind of Garden of Eden, perhaps our earliest home on the planet.

In addition to the monkeys and the antelope, another creature lived there as well. It was mostly apelike but also strikingly humanlike. We know this because researchers have recently found more than a hundred pieces of fossilized bone and teeth. These pieces probably came from at least 30 original individuals. They date back to about 4.4 million years ago (Mya).

Of all the bits of bone that have been found, one individual stands out because of the extraordinary completeness of her fossil skeleton. In the catalog of fossils, she is known as ARA-VP-6/500. But to the scientists who discovered ARA-VP-6/500 and who worked for years reassembling her tiny fragments, she is known as "Ardi," short for her scientific name, *Ardipithecus ramidus*. For scientists in the field and indeed for all of us, Ardi has come to represent her kind, giving us a reconstructed face to look at in wonder as we contemplate the extraordinary story of our human past (White et al. 2009).

When she was discovered in 1994, her fossils were a mess. The bones were poorly fossilized, scattered, crushed, and so fragile they could not be picked up until they were encased in plaster. For fifteen years, an international team of scholars scoured the recovery site for fragments. These were protected and shipped from the field site in Ethiopia's Rift Valley near the town of Aramis to the capitol city, Addis Ababa. There, researchers reassembled her bit by bit, often relying on computer models to figure out the best fit for her tiny bone fragments (Gibbons 2009).

Only in 2009 were researchers ready to publish their results. In the journal *Science*, they introduced Ardi and her kind to world. What they gave us, thanks to their exceptional persistence, was an unparalleled window on our past. In the words of Ann Gibbons in the journal *Science*, Ardi is "a leading lady of human evolution ever since she got top billing on the cover of *Science* in 2009" (Gibbons 2013).

Bones of Contention

From the moment of her scientific debut, Ardi sparked controversy among experts. Where exactly does she fit in the story of human origins? Is she our great-great-grandmother? Not very likely at all, at least not in anything like a literal sense. How, then, does she fit into the human family tree? Does she play a critical role, or is she something of a marginal figure?

Beyond doubt, she is our best window on this particular period in our distant past. Any attempt to look at the period between 4-6 Mya has to acknowledge that Ardi and her kind offer us the most complete picture of what ancestral human life was like back then. Because they go so far back in time, the various forms of *Ardipithecus* offer us the best insight into that extraordinary moment when humans and chimps diverged (Lovejoy et al. 2009).

Before that time, long before today's humans and chimps existed in their present form, there was one large community, ancestral to both humans and chimps. Within that large community, something like two great families began to go separate ways, probably both geographically and biologically. Each great lineage began to follow its own course, evolving and adapting to the environment in distinct ways. One of these families after that divergence is the ancestral community for today's chimpanzees. The other leads to us.

This process of divergence has happened again and again in our past. It is a fundamental process in evolution. By diverging, one population or species can become several different species.

Ardi takes us back in time almost to the era of this great divergence, the last one that has led to two different living species, humans and chimps. The main dispute is whether *Ardipithecus* belongs on the chimp side or the human side of the divergence. Is Ardi on the side that leads to chimps or to us? There is not enough evidence to be conclusive, but the evidence we do have seems to lean in the direction of seeing *Ardipithecus* on our side of the divergence (Harrison 2010; White et al. 2015).

We will come back to this debate a little later. For now, let me tell you more about Ardi.

Our First Leading Lady

Ardi, her discoverers believe, lived until she was about thirty years old. She weighed about 110 pounds and stood almost four feet tall. Her brain was small by our standards, chimp-sized or about one-third the size of ours. Her bones, particularly the pelvis, strongly suggest that when she was on the ground, she walked upright on two feet (Suwa et al. 2009a).

All the great apes today *can* walk upright. But that is not their most graceful way of getting around. To make it work, they prop themselves up with their hands, balancing themselves and supporting the weight of their upper bodies on their knuckles. In order to support this weight, their knuckles are quite different from ours. They also walk with their hips and knees bent. This is called a "bent-hip-bent-knee" gait or just "BHBK" (Lovejoy et al. 2009a).

The knuckle-walking, BHBK gait is built into the anatomical differences between apes and humans. Compared to humans, apes like chimpanzees have different hands, different knees and hips, and different backbones or spines, especially in the lumbar region. In chimps, this lumbar region of the lower back is shorter and more rigid than in humans. That's great if you are hanging from branches but not so good if you are trying to walk upright. Not only do they look pretty awkward on the ground, but their gait is really inefficient in terms of energy use. Compared to chimps, we humans seem almost designed to walk and even to run.

If apes are knuckle-walkers while today's humans can run marathons, were does *Ardipithecus* fit? Somewhere in between. Ardi's wrists and hands were not built for knuckle-walking. That fact suggests that mostly, she walked on two legs in a somewhat more characteristically human way.

The feet of *Ardipithecus*, however, tell a more complicated story. Arid could grasp branches with her somewhat chimp-like feet, making it easy for her to move through trees. Like chimps, she had opposable toes, meaning that her foot looked something like her hand and was good for grasping branches. But her toes were somewhat short and bent like human toes, and her foot was probably more rigid like ours, helping her push off when she walked (Lovejoy et al. 2009a; White et al. 2015).

Ardi could *walk and climb*. Her walking was not as graceful and efficient as ours (Lovejoy et al. 2009b). She has even been described as walking "in a weird way" (Gibbons 2013). Not only did she have an opposable toe, but she also lacked the arched foot that makes our walking highly efficient. But Ardi was no knuckle-walker. Her knees and hips were probably not bent, at least not very much. In the trees she probably walked on four limbs through branches rather than hang or swing from them. She seems to have lived her life on two levels, walking fairly efficiently on the ground between trees and able to climb quickly to safety when she needed to escape danger on the ground.

If Teeth Could Talk

Chimps, especially males, have large, sharp canine teeth that can be used to attack outsiders or threaten other males. The "alpha male," in fact, is usually the one with the most threatening teeth. By scaring off the other males, he is the one that mates most often. Over time, this results in "sexual selection." Not all males reproduce at the same rate. In some species, a few males beat the odds by driving off other males. In other species, female receptivity plays a key role in deciding which males mate and leave the most offspring. Whatever male feature or trait makes them more intimidating to other males or more favored by females can become increasingly exaggerated over time.

Something like this seems to have happened with birds. In many species, the male has more brightly colored feathers than the females. The most likely explanation is that over time, males attracted females with bright feathers. Those with bright feathers were the ones who mated, which meant that more and more brightly colored feathers were selected, becoming quite exaggerated in birds like peacocks. Something similar probably happened with chimps and their large canine teeth. Female chimps either preferred scary teeth or, more likely, males with big teeth scared off other males. Over time, bigger and bigger teeth won.

Ardipithecus teeth are different from chimp teeth in intriguing and important ways. Among the *Ardipithecus*, the male canine teeth seem to be noticeably smaller than what we see in chimps. In *Ardipithecus*, males and females appear to have canine teeth of roughly the same size, while in chimps the difference is pretty dramatic. And chimp canines come with a self-sharpener, a built-in honing mechanism that keeps male canine teeth dangerously sharp. This honing structure is missing in *Ardipithecus*. It seems that after the chimp/human divergence, the line that led to chimps evolved more pronounced male canine teeth while the line that led to us evolved toward the smallish canines we have today (Suwa et al. 2009).

What might this mean? One possibility is that breeding strategies were changing. Rather than scaring off the other males and having all the females, maybe the *Ardipithecus* male was becoming more monogamous, bringing food during pregnancy and sticking around during the rearing of the offspring. In that case, rather than one alpha male scaring the others away, the males may have cooperated in defending and helping to feed the group.

Knowing all that from a few four-million-year old teeth seems like a stretch, but something interesting appears to have been going on. During these critical stages of human evolution, over a span of several million years, ancestral hominins were becoming more socially cohesive, more monogamous, and more invested in their offspring. Males in particular were spending less time fighting each other and more time providing for their mate and their offspring. Some might question whether this transition is fully complete in today's human males.

Another possibility is that in our earliest ancestors, canine teeth became smaller because they were no longer needed to attack prey or to tear meat from bones. Some have suggested that when our ancestors made tools to do these things, canine teeth became smaller.

But Ardi seems to point to a different story. Nothing suggests that *Ardipithecus* made tools. And yet their canines were relatively small and seem to be part of a general pattern over time that leads toward even smaller teeth and less pronounced male canines. The use of tools may have played a role in this process later on. But the change in teeth seems to predate the rise of tools by a million years or more. So what is the explanation?

Owen Lovejoy suggests that canine tooth reduction is best explained by seeing how it is connected with other traits. These traits fit together into a kind of package deal, what he calls an evolutionary "adaptive suite." A suite is a set of traits that complement each other and only make adaptive sense together. In human ancestry, tooth size fits together with reproductive adaptations, bipedality, and brain size (Lovejoy 2009).

Lovejoy thinks it happens like this. Over time, bipedality, unusual reproductive strategies, and increased brain size come into existence together because they complement each other. How? Bipedality means less time in trees and more time on the ground. The group covers greater distances in pursuit of more varied and potentially richer sources of food, including animal sources. But that means more risk, especially for pregnant females or those with young offspring, all the more so if mobility in the trees is diminished. When that happened, males became more specialized at foraging at greater distances, leaving females in a relatively safe place but without complete access to food and somewhat more dependent upon males to provide it (Lovejoy 2009).

One scary alpha male cannot do that for a large troop. Male cooperation is needed, so there is no longer a need for supersized teeth. There is a need, however, for the male to keep a relationship with the female during the increasingly long time that it takes for infants to mature, especially as bigger brains are evolving. If infants need time to mature, their nursing mothers need a consistent supply of food, the richer the better. And so does the newly weaned but still largely helpless young child. For this to happen dependably, the male and female need a strong connection, most likely a bond based in sexual activity. For chimps, the female's time of fertility is highly visible to the troop. It is an advertisement to the alpha male to procreate and sire the next generation. But among human female ancestors, the time of fertility is hidden. No one in the troop could tell when the female is fertile. This means her male can leave to gather food without too great a likelihood that when she becomes pregnant, her offspring would be fathered by another male. He will assume it is his, and he will continue to protect and provide for mother and child.

Teeth alone don't offer this much information. But for the team of researchers that have worked most closely on *Ardipithecus*, the question raised by the smaller canines is best answered by looking at a broad pattern of changes. They suggest that "reduction of male canine size and height, especially of the upper canine, signals a fundamental change in social behavior. Moreover, bipedality and male canine feminization appear to have been evolutionarily coupled" (White et al. 2015).

A complex tale, to be sure, but plausible and at least arguable. If it is anywhere close to the truth about what actually happened, it suggests that various features that distinguish us from other animals come into existence because of the way in which they complement each other. In addition, according to Lovejoy, these changes began millions of years ago, at least as far back as *Ardipithecus*. Lovejoy believes that recent findings "suggest that hominid cerebral evolution extends deep into time (Lovejoy 2009).

Ardipithecus may not show us everything about how these traits fit together. But these fossils do take us back in time, back far enough to shed some light on one of the most consequential moments in our past. This is the moment when our ancestral line diverged from a common ancestor that we shared with the chimpanzees. More than anything else, this human/chimp last common ancestor defines the jumping-off point for the human story, marking the beginning of our kind as a distinct form of life on the planet. What makes us human? Whatever exactly *humanity* is, it begins here.

Ardi and the Last Common Ancestor

Back far enough in time, well before *Ardipithecus* and long before today's chimps or humans existed, there was an ancestral community from which today's humans and chimps evolved. What we have learned about *Ardipithecus* has led some experts to rethink what we know about this common ancestral community, often referred to simply as the "Last Common Ancestor" or LCA. The LCA existed at the point of divergence. It was the ancestral community that separated into at least two lines, each evolving over millions of years and eventually becoming today's chimps and humans (White et al. 2015).

Sometimes people say mistakenly that humans evolved *from chimps*. Experts in evolution never quite saw it that way, but for a long time many of them did think that our ancestors were knuckle-walkers, just like chimps. But now along comes Ardi, walking upright, not exactly like us but more like us than like a chimp. And she takes us back close to the time of the human/chimp LCA. She is our best window on what the LCA was like. For some experts, Ardi forces us to take a new look of our last common ancestor. *Ardipithecus* calls into question the standard view that our common human/ape ancestors were all knuckle-walkers. Because Ardi walked upright, it is more likely that the LCA walked upright.

And if that is true, the line that led to us humans did not start out with knuckle-walking and then lose this trait. We never had it. But that would mean that at the beginning of their lineage, neither did chimps. The LCA moved quite well in the trees but not so well on the ground. Our lineage evolved by walking upright with efficiency and grace while losing agility in the trees. The chimp lineage evolved knuckle-walking for comparatively awkward ground movements while becoming very agile in the trees. Humans and chimps still resemble each other somewhat, but there are significant anatomical differences that underlie our different ways of getting around. Hands, knees, hips, and spines are all different, depending on whether one is a knuckle-walking chimp or a bipedal human (Lovejoy and McCollum 2010).

Owen Lovejoy and Melanie McCollum call attention to a crucial difference in the spine, specifically in the lumbar region and to the absence of flexibility among knuckle-walkers. They claim that "the earliest hominids were able to functionally achieve bipedality because they never rigidified their lumbar spines. Instead, they evolved an opposite morphology—a *reduction* in iliac heights and a *broadening* of the sacrum." They also claim that what we now know about the anatomy of *Ardipithecus* is probably our best clue to the physical structure of the last common chimp/human ancestor. They write: "The generalized structure of earliest hominids that permitted this sequence of events is almost certainly extendable to the LCA" (Lovejoy and McCollum 2010).

But if the LCA was not a knuckle-walker, then a serious problem arises. Chimps must have evolved this trait. And so must have the gorillas, each of them separately. For traditional evolutionary theory, this is a bit of a stretch, so some experts hold to the idea that the LCA was a knuckle-walker. Others point to *Ardipithecus* as evidence that the LCA was no knuckle-walker, so it must be true that the trait evolved at least twice. They appeal to what evolutionary biologists call *convergent evolution* or *homoplasy*, a well-documented process by which many traits are independently evolved.

Did convergent evolution occur here, and is that the explanation for knuckle-walking chimps and gorillas? After all, insects, birds, and bats all independently evolved wings usable for flight. Did chimps and gorillas independently evolve the anatomical features that allow knuckle-walking? If not, then the only alternative is to think that the last common human/chimp ancestor was a knuckle-walker and that *Ardipithecus* lost this trait awfully fast.

Additional support for convergent evolution is provided in a recent study by Tracy Kivell and Daniel Schmitt. They compare the distinctive features of the hands of chimps and gorillas. Yes, both animals are knuckle-walkers, but their knuckle-walking anatomy is sufficiently different to "support a hypothesis of independent evolution of knuckle-walking behavior in the 2 African ape lineages" (Kivell and Schmitt 2009).

Just a few years ago, nearly everyone thought that the LCA must have been a knuckle-walker. The reasoning seemed straightforward. Humans, chimps, and gorillas all share a common ancestor. Two out of three of these species are knuckle-walkers, so the LCA most likely was, too. Even as late as 2009, Kivell and Schmitt argued that their evidence was still too inclusive to decide between two possible options. Did human bipedalism evolve from a terrestrial knuckle-walking ancestor or did knuckle-walking evolve from a more generalized arboreal ancestor? Now they argue that the evidence tilts in favor of the view that human bipedalism is not evolved through a loss of terrestrial knuckle-walking features but by adaptation from life in the trees.

Even more recently, new evidence provides additional support for the idea that the LCA was not a knuckle-walker and that "that knuckle-walking in chimpanzees and gorillas resulted from convergent evolution" (Morimoto et al. 2012). What seems to have happened is that chimps, gorillas, and humans all came from an ancestor that was good in the trees but not so adept at walking on the ground. Each lineage evolved its own solution. Gorillas and chimps separately evolved the ability to support their weight on their knuckles, while the human lineage evolved more and more efficient ways to walk upright. Today, the human foot is highly efficient for walking and running. Increasing efficiency on the ground goes back at least to the time of Ardi.

The picture that is now emerging is that bipedalism goes back 6-7 million years. The LCA was not a knuckle-walker. From the start, our ancestors after the human-chimp divergence would have been climbing and also walking, although not yet very efficiently or gracefully. In the line that led from the LCA to chimps, knuckle-walking evolves, along with greater agility in the trees. In the line that led from the LCA to us, agility in the trees is lost and efficiency on the ground is gained. In simple terms, this suggests that from the time of the chimp-human divergence, the evolution of efficient upright walking upright is a defining feature of the human side of the divergence (Morimoto et al. 2012).

We also know now that there are several different ways to be bipedal. Just as there are at least two ways different ways to engage in knuckle-walking, one evolved by chimps and the other by gorillas, so ancestral humans walked upright but with differences between various parts of the family. Bipedalism is no simple yes-or-no trait. It is ancient, gradual, and diverse in form. And it seems to have existed in several forms side-by-side. An Ardi-style gait, somewhat awkward because of the opposable big toe, seems to have persisted until about 3.4 Mya, long after other forms of human ancestors had lost their opposable toes and were walking more gracefully (Haile-Selassie et al. 2012).

Why is this important? It changes the way we see our pre-human ancestors. When we say we evolved, we do not mean that we evolved *from chimps* or even from creatures that strongly resembled chimps, with traits like knuckle-walking. We cannot look at chimpanzees and say to ourselves: "Once we looked and behaved like that." We have to be really cautious about claiming to know very much about early humans by studying today's chimps. The two are significantly different in many important ways.

Another implication is that one of our most important human features—the ability to walk upright with increasing levels of grace and efficiency—is a feature that goes way back into our pre-human past, back millions of years, back to Ardi and before. Bipedalism may be a key human trait, but over millions of years of human evolution, it has not only changed but has existed in several forms at once.

Before Ardi

It is one thing to put the pieces of a skeleton together, as the researchers who found Ardi have done so admirably. It is another thing to put the skeletons together into a coherent picture of what happened. One problem is that we just do not have enough specimens. That problem gets even worse when we ask the obvious question: What do we know about primate evolution before *Ardipithecus*? We do know something because we have a few findings. But putting the pieces together into a coherent story of our past is a real challenge. Here is a brief summary of what we know.

We human beings belong to the biological order of primates, which includes not only the great apes but also monkeys of all sorts. The first primates were tiny, mouse-like creatures that appeared somewhere around 65 million years ago. Over time, they spread across Asia, Africa, and the Americas. They evolved and diverged into different lines of descent, and their descendants are the monkeys, apes, and human beings that we know today.

Among the primates are the great apes. Biologists call all the great apes by their family name of *hominidae*, which suggests that they are somewhat human-like. Today there are four living types or "genera" of the family *hominidae*: (1) chimpanzees and bonobos, (2) gorillas, (3) humans, and (4) orangutans. This family—*hominidae*—goes back some 15 to 20 million years, when we diverged from other primates. The formal Latin *hominidae* is frequently shortened to "hominids." Strictly speaking, "hominids" refers to all the great apes. Often, however, even top scientists will use "hominids" to speak of the line that leads to humans. Here in this book, except when quoting others, I use the more precise word "hominins," reserving "hominids" for the whole family of apes.

Today, it is sad to say, the entire *hominidae* family is threatened with habitat loss and eventual extinction except for humans.

Over time, the *Hominidae* family evolved and diverged into separate lines of descent, leading to gorillas, orangutans, and chimps/humans. Then about seven million years ago, the chimp/human line diverged once again, separating and leading at last to chimpanzees and humans, closely related but quite distinct.

When we compare human and chimp anatomy, we can see the resemblance. But the best evidence to support the idea that humans and chimps are closely related comes from genetics. Thanks to the Human Genome Project and more recently to the Chimpanzee Genome Project, we know that we humans share more of our DNA with chimps than with any other living creature. In fact, considering just how close the DNA is, we should wonder why our looks and behavior are so different.

When did the human/chimp divergence occur? Until about 2010, experts were faced with two different calculations, one based on analysis of fossils and the other based on genetics. Based on a comparison of the chimp and human genomes, scientists were pretty sure that the divergence occurred somewhere between 4-6 Mya. But some fossil finds suggested earlier dates, more like 7 Mya.

Beginning around 2010, however, experts in genetic evolution used new information to reassess estimates of key events in human origins, such as the date of the human/chimp divergence. These recalculations are likely to continue as new information becomes available, but the latest estimate of dates from genetics is pretty much in agreement with fossil dating. For the date of the human/chimp divergence, some experts offer a range estimate of 3.7 to 6.6 Mya while others push the date back to 7-13 Mya. Greater agreement will likely come with more information and more advanced methods, but a date in the 7-8 Mya, roughly in the middle of current estimates, seems to fit in general terms with the latest fossil discoveries.

So now the best estimate for the date of the human/chimp divergence is approximately 7-8 Mya. It also seems clear this occurred somewhere in east Africa, maybe as far west as modern Chad in north-central Africa. In that time and place, an ancestral community of *hominidae* diverged into what became at least two separate lineages, each evolving in its own way, one becoming increasingly human-like while another gave rise to today's chimps.

Ardi takes us back a long way toward this fateful moment of divergence, to 4.4 million years, to be precise. But if the 7-8 Mya divergence date is correct, Ardi only takes us part of the way. So we cannot help but ask: What came before Ardi? In the past decade or so, researchers have discovered fragments of skeletons, just enough to describe three different species that may be part of our lineage before Ardi.

The first of these is very much like Ardi, enough to be classified in the same genus, *Ardipithecus*. Travelling back in time, now to somewhere between 5.2 and 5.8 Mya, we come to what scientists have labeled *Ardipithecus kadabba*. The older species pre-dates Ardi's kind (*Ardipithecus ramidus*) by about a million years. Even so, the resemblance is strong, enough to put them both in the same genus.

As we go back further to about 6 Mya, we come to a set of findings that some regard as an even earlier form of *Ardipithecus*. Others insist that this is a completely different genus, one they call *Orronin tugenensis*. About 20 different specimens have been collected, just enough to fuel claims and counterclaims about *Orronin*, many of them exaggerated and almost all of them controversial. Brigitte Senut led the team that discovered *Orronin*. She insists that this creature is earlier that Ardi and yet more human-like. If she is right, then *Ardipithecus* is not a human ancestor, but *Orronin* may very well be (Senut et al. 2001). A more recent and cautious assessment has been published by Brian Richmond and William Jungers. They suggest that *Orronin* is closely related to *Ardipithecus* and seems to fit what might be expected for its era, roughly at a point between the LCA and Ardi (Richmond and Jungers 2008).

Earliest of all and perhaps the most intriguing find is an approximately 7 million year old skull reported in 2002. A fairly complete cranium and a few other fragments have been found. Michel Brunet, who led the team that made the discovery, named it *Sahelanthropus tchadensis* for the African region (the Sahel) and the modern nation of Chad where it was found. Two things make *Sahelanthropus* really interesting. The first is its location. Chad is not in East Africa but in the north central part of the continent, more than 1600 miles west of the more commonly searched sites in Ethiopia and Kenya. That suggests that whatever might have been happening in terms of human origins, it was happening over a very wide expanse, implying that there are many other undiscovered sites just waiting for productive field work (Brunet et al. 2002).

The second intriguing thing about *Sahelanthropus tchadensis* is its date of almost 7 million years. This date takes us back to the very beginning of the human line, right at the moment of human/chimp divergence. Unfortunately, we do not know much about this species. Its environment, like Ardi's, was probably open woodlands. Brunet argues that *S. tchadensis* was habitually bipedal, surely able to move through the trees with ease but more adept at upright walking on the ground than modern chimps.

Brunet's team was able to reassemble a fairly complete skull. The brain was small, only slightly larger than the typical chimp's brain. Based on their analysis, they argue that *S. tchadensis* has cranial features that resemble later hominins. They suggest this shows that it is probably on the human side of the human/chimp divergence. According to Brunet, these features "clearly illustrate its hominid affinities temporally close to the last common ancestor of chimpanzees and humans, and also that it cannot be related to chimpanzees or gorillas" (Brunet et al. 2002).

Evolving Debates

Not everyone agrees with Brunet. Some, in fact, object pretty strongly not just to the claim that *S. tchadensis* lies on our side of the divergence but to the very idea that the human family tree is subject to revision in light of each new finding. Haven't we learned anything definitive or final about human origins after decades of research, they ask? Is it really all up for grabs?

On one side are those who argue that *S. tchadensis*, *O. tugenensis*, and even *Ar. ramidus* cannot be classified with any certainty as having been on the human side of the chimp/human divergence. One of their objections has to do with timing. These creatures lived too far in the past, too close to the time of the LCA, to allow for clear classification. They have a point. The closer something lies to a dividing line, the harder it is to know on which side of the line it belongs. Now, with more recent calculations of the time of divergence based on genetics, their argument seems not to work so well, especially with *Ardipithecus*. In the case of *S. tchadensis*, however, it is still a fair point to make. This form of ancestral life lies so close to the LCA that it is impossible, based on what we know now, to decide for sure on which side of the divergence it belongs. Even Brunet agrees that *tchadensis* is "temporally close" to the LCA .

The second objection takes us to a more profound question, not just about whether Ardi belongs in the human family tree, but about the limits of our knowledge of our earliest lineage. How much do we really know about this period in our past? Do we know enough to claim the right to construct even a rough draft of the story of our emergence after the LCA? Some, like Brunet, argue that the scientific quest for human origins is still a new science, and that today's provisional notions are sure to give way to what we will learn from new findings. Others argue that the field is pretty solid and not so likely to change.

For example, Estaban Sarmiento writes this: "Human evolutionary studies are not a new science where every new find revolutionizes interpretations of our past" (Sarmiento 2010). Brunet argues back by pointing out that the whole idea of collecting and classifying fossil humans is "quite recent." He comments: "As palaeontologists and palaeoanthropologists, we have always to remember that our interpretations have, at most, a life expectancy that usually does not go beyond the next new major fossil discovery." Finally, Brunet adds this: "During the past 150 years, most of the models for hominid evolution have been overturned by successive discoveries. This fact obviously highlights the importance of fieldwork" (Brunet 2010). To which Sarmiento replies: "A purported fossil ancestor that must overturn nearly all we know about our evolution to fit into our lineage is unlikely to be such an ancestor" (Sarmiento 2010).

Bernard Wood and Terry Harrison also urge caution in making major revisions and additions to our human family tree. They write this: "…we do not claim that *Ar. ramidus*, *S. tchadensis* and *O. tugenensis* are definitely not hominins. We do, however, advocate that those palaeoanthropologists whose considerable and much valued efforts in the field are rewarded with fossils as significant as those from Aramis, Toros Menalla, Lukeino and Malapa acknowledge the potential shortcomings of their data when it comes to generating hypotheses about relationships. We urge researchers, teachers and students to consider the published phylogenetic interpretations of these taxa as among a number of possible interpretations of the evidence" (Wood and Harrison 2011).

The point here is not that these researchers are being careless or rushing to conclusions but that the task before them is difficult. Fitting the fossils together to form a coherent story is not a simple task. Discovering more fossils will help, but more discoveries will certainly fuel more debates. Some critics of evolution see debates among evolutionary biologists as counting against its credibility as a science. That is backwards. Vigorous debate is evidence of rigorous work and rapid advances. It is not a weakness but a strength of the scientific enterprise that researchers hold their theories as "possible interpretations."

Even if the precise story is debated, there is no question among these researchers about the general story line. Our human lineage diverges from the chimp lineage about 7 million years ago, gradually evolving by diverging and adapting to new environments, through the era of *Ardipithecus* and then to the age of *Australopithecus*, the next chapter in the story of human origins.

3

Lucy and the Next Two Million Years of the Human Story

When "Lucy" was discovered in 1974, she was an instant celebrity. Her fragile bones were laid out in careful arrangement and photographed for all the world to see. Nothing so ancient and so complete had ever been found before. She became the icon for all things pre-human, instantly connecting us as never before to our distant ancestry. Speaking somehow from the silence of her bones, she haunts us with a distant echo of our past.

Her brief life was spent in modern day Kenya about 3.2 Mya. She is the defining example of the genus *Australopithecus*, widely believed to be the genus that sets the stage for our own genus, *Homo*. Lucy's genus of *Australopithecus* included several forms or species, and Lucy's own type or species—*Australopithecus afarensis*—is by far the best known (Kimbel and Delezene 2009).

Back in 1974 when the world first learned about Lucy, no one knew about Ardi. Now that we know about *Ardipithecus*, we can't help wondering if *Ardipithecus* gave rise to *Australopithecus*. Something like that seems plausible, given the timing and the location of all these remains. The details of the evolutionary story, however, are still being debated and key discoveries are still waiting to be made. We do know that Ardi and her kind disappeared from the forests of East Africa about 4.3 Mya. As *Ardipithecus* departs, *Australopithecus* arrives, ushering in a new stage in the story of our origins.

If there were a simple story line here, it would go like this: Ardi's genus appears about 5.8 Mya and disappears about 4.3 Mya, yielding the stage to Lucy, whose genus appears about 4.2 Mya. They vanish when our own genus, *Homo*, is beginning to appear more than 2 Mya. But the reality, we are learning, is anything but simple. There is no sequential story line, as if our ancestral forms were all in a neat line, each walking across the east African evolutionary thoroughfare one after another, like high school bands at a parade. At times they shared the stage with each other, overlapping each other's species lifespan, apparently coexisting in various forms at once. The more we learn about our ancestors in this period of 3-4 Mya, the more difficult it becomes for experts to construct a neatly sequential story of how these ancestral forms fit together (Kimbel and Delezene 2009).

Even with all our unanswered questions about the broad patterns, we actually do know a lot about Lucy and her genus. This chapter is her story and the story of her extended family, a portrait of the distant human predecessors that we call *Australopithecus*. As a genus, their timespan is more than 2 million years, roughly from 4.2 to 2.0 Mya, taking us to the arrival of the genus *Homo*. The bones of *Australopithecus* urge us to ask: Who were we before we became human?

Lucy and her Family

Lucy rocked the world of paleoanthropology when she was discovered in 1974 mainly because of the completeness of her skeleton. Not that she came entirely out of the blue. Fifty years earlier Raymond Dart announced the discovery of what he called "the Man-Ape of South Africa" or *Australopithecus africanus*. At the time, Dart's announcement was greeted with uncertainty and skepticism, even derision. His discovery of the fossilized skull of a child, now known as the "Taung Child," launched modern paleoanthropology. The ancient skull revealed a small brain but also small, somewhat "human" teeth.

In the 1920s and 30s, many experts argued that Dart's discovery was just another ape. Everyone knows, they said, that the first humans evolved in Asia, not Africa. Why should we even look in Africa? It took a few decades for Dart's views to be accepted. Other small discoveries we made, setting the stage but not fully preparing either the experts or the public for Lucy. Suddenly the experts had some way of putting the smaller discoveries into an anatomically coherent pattern. Individual teeth and bones are helpful finds, but there is nothing like an even partly complete skeleton to help provide critical perspective into the overall anatomy and the proportions between arms and legs, for example. At about 40% complete, Lucy became something of the Rosetta Stone of paleoanthropology, helping experts decipher other finds.

One of the first things everyone noticed about Lucy is that she was small. She was only about 37 inches or 1.1 m tall. If Lucy is the prime example of *Australopithecus*, maybe they are all about this size. Or was Lucy unusual? Or maybe males were significantly bigger than females, yet another example of the sexual dimorphism that is common among primates. In 1974, there was no way to answer these questions. More recent discoveries now seem to show that Lucy was indeed unusually small when compared to other examples of *Australopithecus* (Kimbel and Delezene 2009; Reno and Lovejoy 2013).

For example, a fairly complete skeleton was discovered by Ethiopian scientist Yohannes Haile-Selassie in 2005. The skeleton's catalog name is KSD-VP-1/1, probably a male, and he lived 3.58 Mya. His finders called him Kadanuumuu, which means "big man" in the local language. At times he is simply referred to as "Lucy's Big Brother." Like Lucy, KSD-VP-1/1 is a remarkably complete skeleton. He was about 30% bigger than Lucy, probably reaching a height of 5'2." Even accounting for the likelihood that males were at least somewhat larger than females, he is bigger than expected, suggesting that Lucy may have been unusually small for her kind. Lucy's big brother also walked upright and shows signs of spending little time in the trees. According to Haile-Selassie, this individual was "fundamentally similar in morphology to AL 288-1," meaning that he had the same general proportions as Lucy. But Kadanuumuu dwarfed Lucy, and he lived about 400,000 years earlier (Haile-Selassie et al. 2010).

Given their differences, do Lucy and Kadanuumuu really belong to the same species? With their size gap and the hundreds of thousands of years that separated them, is it right to see them both as *A. afarensis*? Haile-Selassie thinks so. He sees their differences as simply reflecting the passage of time, as representing early and late forms within an evolving species, but not as separate species. They are members of the same "chronospecies," slightly different forms of the same species over time, enough like Lucy to be seen as part of her extended family over time.

If Lucy had a big brother, she now has a little sister as well, called "Selam" by those who found her. Her discovery was reported in 2006. Selam is a relatively complete fossil of a young female *Australopithecus*, thought to be approximately three-years-old. She was found near Dikiki, Ethiopia. Her fossil remains were embedded in stone, and the removal process is taking years to complete. The reward for all the hard work is the most complete *A. afarensis* remains yet discovered. As her bones are revealed, so are her secrets (Alemseged et al. 2006).

Selam dates to about 3.3 Mya, just a little before Lucy. What makes her so valuable to science is that she shows how these ancient human ancestors developed through childhood. In 2012, David Green and Zeresenay Alemseged reported that based on their analysis, Selam walked upright and probably had hips and knees that closely resemble our own, but her arms and shoulders were more ape-like than human. That suggests that she was probably a pretty good climber. Just how much time she and her family spent in the trees is a matter of debate. She seems to provide new evidence that hanging from branches was still a common practice. Some experts, however, think that body changes just had not yet caught up with behavioral changes. She spent most of her time on the ground, or would have as an adult, they argue, even though her young body showed signs of spending time in the safety of the trees.

Linking Selam's anatomy to her behavior is a special challenge because she is so young. As we develop from infant to adult, our bodies undergo changes in form and not just in size. What might Selam's body have been like as an adult? Green and Alemseged claim that "growth of the *A. afarensis* shoulder may have followed a developmental trajectory more like that of African apes than modern humans." In terms of behavior, they suggest that this means that although *A. afarensis* walked upright in a way that approximates our own gait, they were also quite capable of climbing and probably spent a significant amount of time in the trees. The researchers conclude with this claim: "The apelike appearance of the most complete *A. afarensis* scapulae strengthens the hypothesis that these hominins participated in a behavioral strategy that incorporated a considerable amount of arboreal behaviors in addition to bipedal locomotion" (Green and Alemseged 2012). In other words, Selam's family could walk but they probably could swing or at least hang from the branches pretty well.

Next of Kin

Lucy, Kadanuumuu, and Selam may be the best-known representatives of *A. afarensis*. But they are not alone in telling us what life was like 2-4 Mya. Today we have more than 400 specimens of teeth and bones from *A. afarensis*. We also know about other forms or species of *Australopithecus*, and more examples are being added almost yearly.

For example, we know about what is probably the earliest form of *Australopithecus*, a somewhat shadowy and controversial form that dates to about 4.2 million years ago. It is called *A. anamensis*, known from about 100 fossils that offer only glimpses at what these creatures were actually like. Some argue that *A. anamensis* is the evolutionary source for other forms of *Australopithecus* (White et al. 2006). They were quite a bit like Ardi in terms of brain size. They also walked upright more efficiently than any living ape (Kimbel et al. 2006).

If there are too few fossils from the period between 4.2 and 3.6 Mya, the opposite is true when it comes to the more recent end of the period of *Australopithecus*. In 2011, researchers in Ethiopia led by Haile-Selassie found three fossilized jaw fragments dating from 3.3 to 3.5 Mya. The new jaws are somewhat more robust and have heavier teeth enamel than *A. afarensis*, suggesting different diets. The team's analysis, published in 2015, led them to claim that they found a new species, which they called *Australopithecus deyiremeda*. This announcement came on top of other similar claims for other variants of *Australopithecus*. Experts agree that there is a significant amount of variation in these samples. Some, however, argue that the very concept of a species involves variation within a species. *A. afarensis*, they believe, is a species with unusual longevity and variation (Haile-Selassie et al. 2015; Kimbel and Delezene 2009).

Others counter by claiming that multiple species of *Australopithecus* may have existed side by side in the ancient world of East Africa, along with *A. africanus* living further south. The wide variation in teeth and bones is best understood by assigning them to more than one species, perhaps even going back and reassigning fossils already classified as *A. afarensis*. The magazine *New Scientist* quotes paleoanthropologist John Hawks as saying: "If Haile-Selassie is right, I think it's only reasonable to conclude that some unknown number of *Australopithecus afarensis* skeletal remains actually belong to this new species instead. This means that everything that has been written about variation, function and the anatomy of *Australopithecus afarensis* from fragmentary remains must now be in doubt" (Sarchet 2015).

Despite the variability *within Australopithecus*, there is one thing that sets off *Australopithecus* from Ardi. It is the loss of Ardi's opposable toe, so useful for holding on to tree branches. That is not to say that *Australopithecus* did not climb trees and hang from branches. During this period from 4.2 to 2.0 Mya, human ancestors were becoming more and more adept at living on the ground. Some of the most striking evidence of their ability to walk is a set of footprints that date to 3.66 Mya, left by a group of *Australopiths* at Laetoli in Tanzania. Researchers have compared these ancient footprints with the prints left by humans walking today under similar circumstances of wet sand. Their conclusion is that whoever left these early footprints had a human-like gait (Crompton et al. 2012). This is yet another confirmation that early on, maybe all the way back to the last common ancestor, the ancestors of humans were not knuckle-walkers (Raichlen et al. 2008).

Toes and Toddlers

With the passage of time, our ancestors were slowly losing the ability to move quickly and safely through the branches. It is not just that the opposable toe is gone. Shoulder strength and spinal flexibility were changing. Compared to Ardi, Lucy is less able to climb than to walk. Ardi walked upright, but with her opposable toes, she could climb to safety in the trees even while carrying an infant. Lucy probably spent a fair amount of time in trees, foraging or escaping danger below. But because of a number of anatomical changes slowly accruing over time, climbing was harder for *Australopithecus* than for *Ardipithecus*.

This was especially true for new mothers. In addition to the loss of a grasping foot, another reason was coming into play. Over time, Lucy's *Australopithecus* kin were giving birth to larger offspring with bigger brains. This poses several new problems. Compared to her predecessor (and even more when compared with chimps), the *Australopithecus* mother is less able to take to the trees for safety while carrying her young. Her newborn is relatively harder to carry. Climbing is especially difficult grasping branches with one hand and a child with the other (de Silva 2011). All this seems to suggest that she needed help feeding and defending herself if she and her infants were to survive. Apparently, she got what she needed, most likely from a pair-bonded male who provided food and protection, probably in cooperation with other males.

When we compare chimps and humans, we see some obvious differences in toes, teeth, and procreative strategies. Chimps are sexually promiscuous while humans, for the most part, form sexual pair-bonds. Some experts, such as Owen Lovejoy, argue that the key trend-line was already established more than 4 Mya with Ardi. Already at that time, he argues, we can see the rise of a set of traits that include walking upright, pair bonding, extended infancy, and eventually larger brains (Lovejoy 2009). Not all agree that this trend is present so far back, but nearly everyone recognizes that it is well underway by the time of Lucy and *Australopithecus*.

Our ancestral mothers faced a two-fold evolutionary challenge. Larger brains and bipedalism might seem like benefits, but for the ancestral human mother, both pose life-threatening challenges. They faced the combined challenge of heavier infants with nutritionally needier brains and, at the same time, were less able to climb, especially while carrying a child. Without a nurturing and protective community, they and their infants would have been easy prey.

This is why some experts conjecture that the changes that we can see in the fossils must have been accompanied by changes that we cannot discover directly by studying old bones or teeth. Some argue that in order to survive, the *Australopithecus* mother must have been able to form a somewhat stable pair bond with one male who cooperated with other males more than he competed. For this to happen, they suggest, the female's ovulation and time of fertility is hidden, meaning that males have no way to know when to mate in order to produce offspring. In addition, mammary glands are permanently enlarged, meaning that the male could not tell when she was no longer nursing and therefore ready to conceive again. These changes helped support pair bonding as a reproductive strategy.

Of all the changes that come in the wake of pair bonding, by far the most momentous is the eventual growth in brain size. Care for the mother and infant meant that it was possible in time for humans to evolve brains that are larger both before and after birth. But this complex set of changes also meant that it was increasingly *necessary* to have a larger brain. Society was more complex. Pair bonding creates the possibility of deception. "These forces selected for more and more effective strategies (including deception, manipulation, alliance formation, exploitation of the expertise of others, etc.)" (Gavrilets 2012).

No one can claim to see the behavioral changes going on here, much less how all the little changes may have fit together in a whole package. As we learn more of the details of the small changes, we can get a glimpse of how tiny changes in teeth or bones may have played a big role in evolution. For example, careful analysis of the 2011 discovery in Ethiopia of a fossilized bone dating to 3.2 Mya shows clear signs of an arched foot, a marvel of evolutionary engineering that both cushions and adds a spring to our step. The arch is not present in Ardi and probably not in the earliest forms of *Australopithecus*, but it is clearly present by 3.2 Mya (Ward et al. 2011).

Another change that occurs at about the same time is an expansion of the diet. Using carbon isotopic data, researchers have found that *A. afarensis* ate a wider variety of foods than their predecessors did. They began to draw on foods that grow in open grasslands or savannahs or along marshes. The change seems to have occurred between *A.*

anamensis and *A. afarensis*. Perhaps there was a change in climate or in food sources, but the evidence hints at least that the change was in the evolving community rather than in a changing environment. In spite of the fact that these foods had been available a million or so years earlier in the time of Ardi, it was only with *A. afarensis* that they become part of our ancestral diet. According to the researchers, *A. afarensis* marks a "transformational stage in our ecological history, during which hominins in eastern Africa began to expand their dietary resource base to include. . .foods that had been abundant in their environments for at least 1 million years" (Wynn et al. 2013).

As a result of changes in their diet, our ancestors were able to draw on more sources of food. More than that, this change signals greater flexibility and adaptability to a range of environments. When combined with an enhanced ability to walk efficiently, these changes suggest an opening up to new and expanded horizons. Each one of these changes is integral to our definition of what it means to be human. They seem to have come into existence as a kind of package, incrementally, with each change reinforcing the others by making them both possible and advantageous. Each part of the package—efficient bipedalism, diversified diet, pair bonding, reduced canine teeth in males, reduced male/female dimorphism, increased male cooperation, and eventually larger brains—arose over time. The changes interact with each other and with the widening environmental niche that our ancestors began to fill. Each change alone is relatively insignificant or unsustainable. Arising together, they make each other possible and reinforce each other.

The Last Pre-Human

After a run of about two million years, the genus *Australopithecus* seems to disappear from East Africa by about 2 Mya. *A. afarensis*, Lucy's species, seems to disappear even earlier, around 2.9 Mya. Many experts agree that *afarensis* is closely related to the earliest members of the genus *Homo*. But a direct ancestor? The story is probably much more complicated than that. The claim for the earliest evidence of *Homo* is now between 2.75 and 2.8 Mya, based on analysis of a fossil jaw found in Ethiopia in 2013 (Villmoare et al. 2015). It might be tempting to say that in the critical period just before 2.8 Mya, *A. afarensis* gives rise to *Homo*. Even if that is close to the truth, it is far too simple a story to fit all the facts.

For one thing, *A. afarensis* may vanish around 2.9 Mya, but other forms of *Australopithecus* live on for almost another million years. These other forms of *Australopithecus* are not found in places like Ethiopia but in places further south in Africa. The most significant of these is *A. africanus*, Raymond Dart's 1924 bombshell. In the decades just before Lucy was discovered in 1974, nearly everyone thought that *africanus* was our direct ancestor. Because of Lucy, that honor was given to *afarensis*.

Recent discoveries, however, have reawakened old debates. At stake in the debate is nothing less than the correct view of the key step in our past. Where did humanity come from? The most widely held view is that Lucy and her *afarensis* family are our direct human ancestors. The competing view is that Lucy is something of a sideshow while *A. africanus* is in fact a direct ancestor. This argument has taken a new form with the discovery in South Africa of *Australopithecus sediba*. Might *A. sediba* be the transition from *A. africanus* to *Homo*? (Berger et al. 2010).

Most of the evidence seems to argue against this. For one thing, the first appearance of *Homo* seems to predate the appearance of *A. sediba*—odd, to put it mildly, if *sediba* is the ancestor of *Homo*. Another is location. The earliest examples of *Homo* come from places like Ethiopia in East Africa, the same general location as *A. afarensis*. Now we know, however, about the great diversity included in the species *A. afarensis*, as well as the distinct possibility that more than one *Australopithecus* species existed at the same time in the same region. So even if the evidence seems to suggest that *A. afarensis* is ancestral to *Homo*, the pathways of that ancestry are probably complex.

Going against the majority view is the argument that the genus *Homo* arises from the forms of *Australopithecus* that lived in the southern part of Africa. A key part of the argument hinges on the recent discovery of *A. sediba*. During the summer of 2008, Lee Berger was making an initial survey of a cave site at Malapa in northern South Africa. His nine-year-old son, Matthew, came along for the walk and made the key first discovery. What Matthew Berger discovered has raised profound new questions, shaking up the established view of what might have been going on in the human family tree about 2 Mya (Berger et al. 2010).

The Malapa site was rich in fossils from several individuals, including two remarkably complete skeletons. One individual was a juvenile and the other an adult female. Known as MH1 and MH2, these skeletons are far more complete than anything else from the distant past. For example, the adult female remains include a heel, ankle, knee, hip, and lower back or lumbar region. By comparison, the Lucy skeleton only includes the ankle and the hip. One explanation for the completeness of these skeletons is that these individuals were probably encased in mud and falling stone almost at the time of death. This hardened to stone and the bones fossilized, awaiting discovery and then the difficult task of being set free from their stone casing.

As the stone casing is removed, Berger and his team are publishing their analysis of *A. sediba*. Other fossils found nearby are also being prepared and analyzed. All this suggests that there is much more to learn from discoveries at this site. What we already know is provocative and controversial. Who was *sediba* and how does this form of early human life fit into our family tree?

Their brains were small, about 420 cc, about the same as Lucy's, even though *sediba* lived more than a million years earlier than *afarensis*. If brain size is the critical factor, then *sediba* seems best classified as *Australopithecus*, like Lucy and other small-brained ancestors. In other respects, however, *sediba* seems more modern than Lucy. At the same time, overall *sediba* seems to fit less well with Lucy than with the more southern *A. africanus* (Berger et al. 2010). In other words, it seems to lie on a path from *africanus* to *sediba* to *Homo*, not on a path from *afarensis* to *sediba* to *Homo*. Quite simply, *sediba* does not seem to fit where most experts would have expected it to fit (Berger 2012; Berger 2013a).

The central question is whether *A. sediba* is a direct human ancestor (Pickering et al. 2011). Most experts think the answer must be no. Their reason is simple: *sediba* is too late to be an ancestor. *Homo* is already in existence before *sediba* appears. The Malapa bones date to 1.98 Mya, but the most recent evidence suggests that *Homo* is arriving as early as 2.8 Mya. Lee Berger rejects this argument. He is not saying that *sediba* must be a direct human ancestor. But he thinks it is possible. For one thing, just because the Malapa fossils date to 1.98 Mya does not mean that *sediba* only comes into existence at that time. It could be that *sediba* arose long before this date and that Malapa is just a late-surviving remnant (Berger 2012).

Berger's main argument, however, has to do with the questionable quality of the evidence for the first appearance of *Homo*. He questions whether the evidence is clear enough and strong enough to prove that *Homo* goes back before 1.9 Mya. As Berger puts it, claiming to have proof of the appearance of *Homo* is an extraordinary claim, and "extraordinary claims require extraordinary evidence" (Berger 2012). None of the discoveries that are cited as evidence of *Homo* before 1.9 Mya are really very solid, Berger insists.

Most experts disagree with Berger. If they are right and if *Homo* first appears more than 2.75 Mya, then *sediba* (at least in the form of the Malapa community) is not an ancestor to *Homo*. So the most commonly held view is that *Homo* arises from *afarensis*. The key location for the transition is Ethiopia or perhaps Kenya, and perhaps the key transitional form is the recently discovered *Australopithecus gahri*.

Searching for Surprises

In the local Afar language, "gahri" means "surprise." *Australopithecus gahri*, found in 1990 and first described in 1999, is just one of many surprises coming from recent decades of field exploration. In response to its discovery, Bernard Wood said in a news report that "this won't be the last 'surprise'" (Culotta 1999). For those who reject the idea that *A. africanus* is a direct human ancestor, *A. gahri* is just what is needed in the way of supporting evidence. *A. gahri* lived 2.5 Mya, right in the middle of the timespan for *A. africanus*. If we take a linear view of human ancestry, they cannot both be ancestors. Overall, the evidence supports the idea of a transition from *A. afarensis* to *A. gahri* to *Homo*. *A. gahri* undercuts the claim that *A. africanus* and *A. sediba* are any kind of direct human ancestors. According to *gahri's* discoverers, "It is in the right place, at the right time, to be the ancestor of early

Homo, however defined. Nothing about its morphology would preclude it from occupying this position" (Asfaw et al. 1999).

The reconstructed face of *A. garhi* shows both apelike and human features. Nearby bones of large animals dating from the same 2.5 Mya period show signs of butchering using stone tools. No stone tools have been found at the site, but no good tool-making materials are present nearby, so one possibility is that *A. garhi* was making and keeping tools, transporting them as they traveled (de Heinzelin et al. 1999).

The discovery of *A. garhi* raises two important, unanswered questions. First, is *garhi* a direct ancestor of our genus, a kind of bridge between Lucy and us? Most experts tend to think this makes sense. Second, if it is, should *garhi* be classified in the genus *Australopithecus* or in the genus *Homo*? If tool use is the decisive factor, then perhaps *garhi* is human. If brain size is key, *garhi* is an *Australopith* like Lucy. Their brains are about 450 cc, barely larger than chimp brains.

So which is it? Is *garhi* the last *Australopith* or the first *Homo*? Even the discoverers do not claim to know the right answer to this question. "If *A. garhi* proves to be the exclusive ancestor of the *Homo* clade…a cladistics classification might assign it to the genus *Homo*" (Asfaw et al. 1999). Precisely because *garhi* lived so close to the transition between *Australopithecus* to *Homo*, it is not clear on which side of the transition to put it. Those who are looking for clear lines or sharp distinctions are not likely to find them. Is *garhi* the last prehuman or the first human? As uncomfortable as it might make us, the answer to this simple question is anything but simple.

One reason for the complexity is that it is hard to know how to fit the fossils from East Africa together with those from the south, especially now that we know about *A. sediba*. According to Lee Berger, "At first glance *A. sediba* appears to add despairing complexity to our present understanding of the emergence of early *Homo* by adding yet another species, this time with an unexpected mosaic of primitive and derived characters, to what we thought we knew of the experiments occurring between the last australopiths and the first definitive member of the genus *Homo* (c2.0 Ma)" (Berger 2012).

Do we know enough to rule out one scenario or the other? Perhaps there is some truth in both. Why should it surprise us to learn that key evolutionary transitions occur in multiple locations and the results converge later as the result of interbreeding? This is a pattern we will see later in the story of the genus *Homo*. There is no reason to think this process should not play a role in the emergence of our genus in the first place. Perhaps Berger's *despairing complexity* is actually telling us something important about our origins and our existence, something that might better be called *amazing* or *wondrous complexity*.

4
The Dawn of Technology

Once we thought that only humans make tools. But in the 1960s, Jane Goodall discovered that chimpanzees not only use tools but actually make them. Chimps were already known to use stones to crack nuts. But Goodall observed chimps take small twigs, break them to the right length, strip off their leaves, carry them some distance, and use them to fish termites from their mounds. Then, before the unsuspecting termite could get away, the chimps raised the stick-tool to their mouths to enjoy one of their favorite treats. Except for eating termites, we can imagine ourselves acting in pretty much the same way.

Even more recently, other animals such as New Caledonian Crows have been seen making tools. While nothing quite compares to the tools that we humans make, tool-making itself is not unique to humans. As a complex set of capacities, one form of tool-making or another seems to have evolved several times during evolution. The ability to make tools is a competitive advantage.

There is growing evidence to show that our pre-human ancestors made and used tools. The latest findings seem to show that the first stone tools predate the genus *Homo*. Our genus appeared some 2.8 Mya. By then, our *Australopithecus* predecessors had already been making tools for about half a million years. Or perhaps it was some other creature.

No one knows for sure who or what made the first stone tools at about 3.3 Mya. Someone figured out that by striking one stone against another, it is possible to create sharp edges useful for cutting or slashing meat from large animals.

In a way, the story of the first stone tools is the story of the transition from *Australopithecus* to *Homo*. By our standards, the first stone tools are simple. Their effect on human evolution, however, is profound. Tools offer access to new sources of food. Their creation and use place new demands on evolving brains and hands, demands that the tools themselves helped to meet. They become part of the extended environmental niche in which evolution continues. At the same time, these first handheld stone tools became an extension of the hominin body itself. In many ways, tools are at least partly responsible for future human evolution. Before we became human, we made tools. And almost immediately, our tools began to make us more and more human.

Cores and Flakes

In 2010, researchers reported finding what looked like slash marks on bones of large animals, just the kind of thing we might expect to see if someone used stone tools to cut the animal apart. The bones were found in Dikiki, Ethiopia, and were dated to 3.39 Mya. The researchers looked for the tools that made the slash-marks, but they were nowhere to be found. That means that other explanations for the marks cannot be ruled out. Maybe the teeth of animals created them, or maybe these bones were scratched when they were stepped on (McPherron et al. 2010).

Then in 2015, a team of researchers of researchers led by Sonia Harmand described how they literally stumbled across a tool-making location at Lomekwi archeological site near Lake Turkana, Kenya. They found dozens of stones and flakes. "One key surface find was a small rock flake, which fitted in a gap in a buried core as snugly as a jigsaw puzzle piece, confirming that the tools were made through a flaking process," Harmand said at the 2015 conference where the discoveries were reported. One stone weighed 15kg or more than 30 pounds and showed evidence of being used as an anvil.

Compared to the tools made 700,000 years later, the tools found at Lomekwi are unexpectedly large. Because of differences in techniques and mainly because of the huge gap in time, the research team suggests that the tools from Lomekwi, and any other future discoveries that resemble them, be called "Lomekwian." In their article, they write: "we propose for it the name 'Lomekwian', which predates the Oldowan by 700,000 years and marks a new beginning to the known archaeological record" (Harmand et al. 2015).

Prior to the 2015 report, Oldowan tools were thought to be the oldest. The first Oldowan tools are dated to 2.5 and 2.6 Mya, about 700,000 years after Lomekwian. The oldest ones were discovered in Gona, Ethiopia. They resembled tools discovered earlier at the Olduvai Gorge in Kenya, so researchers tend to put them all in the same category in terms of the basic technique and level of skill. They are usually called "Oldowan," after Olduvai Gorge. Since then, so-called "Oldowan" tools like this have been found throughout much of Africa and Eurasia.

Calling them "Oldowan" is probably not the best term, but it is often used as a kind of short-hand to describe what was once thought to be the simplest or first stage of tool-making. Labels like "Oldowan" can mislead us into thinking that there was such a thing as an Oldowan stone tool-making culture or that those who made tools like this shared other common cultural patterns. To avoid this confusion, some scholars think these terms should be dropped altogether. Here is how John Shea objects to using these labels: "This practice reflects an outdated and unrealistic view that these named industries are substantially analogous to named archaeological and ethnographic 'cultures'. They are not" (Shea 2010). Now, however, the discoverers of the tools at Lomekwi have added to the practices of naming a tool-type after the location where they are first found. So now we have Lomekwian, Oldowan, and (as we will see in a minute) Acheulian, each reflecting a different level or mode of tool-making. Experts in the field tend to like these terms. With Shea's warning in mind, we will use them here.

"Oldowan" describes the distinct features of stone tools made in a certain way. The first step is to select a suitable pebble, about the size of a slightly flattened tennis ball. Then this pebble core is "reduced" when it is struck along the edges by a larger stone, a kind of hammer. When it is struck, the pebble cracks so that flakes fall off, leaving the core intact but producing sharp edges that can be used to cut and scrape large animal carcasses or other things.

If they were used to butcher large animals, these stone tools would have had a direct impact on human nutrition. More meat would have become available, along with highly nutritious bone marrow. Those who foraged would have been able to cut a carcass into pieces small enough to carry for long distances to a place of safety or shelter.

Who exactly made these first tools? This is not easy to answer. Once it was thought that some form of the genus *Homo* must have been responsible. After all, only humans or members of *Homo* can make tools. Now it is pretty clear that this is not true. Tools predate the genus *Homo* by half a million years. We cannot say for sure who made these first tools. Oldowan tools, appearing at about 2.6 Mya, were most likely made by *Australopithecus afarensis*. But with Lomekwian tools, the picture is anything but clear. *A. afarensis* lived at 3.3 Mya, but there is no evidence that they lived at Lomekwi. Another species of hominin, called *Kenyanthropus platyops*, was in the neighborhood at the right time.

At the moment, however, no one knows who made these first tools. Does it really matter what name we give to the first hominin engineers? Not really. These labels do not exist in nature. They are modern human distinctions that experts use to categorize their findings. The key point is that around the time these tools appear, our ancestors were changing in many ways. Feet, teeth, brains, hands, diet, and behavior were all undergoing changes, all connected in one way or another with the making and using of tools.

The real question here is how to understand the connection between early tools and all these other changes. Did biological or anatomical changes come first, bringing bigger brains and more coordinated hands, perhaps made possible by less need for mobility in the trees? Or did tool use come first, offering better nutrition and resulting in biological changes? Did humans make the first tools, or did tools make the first tool-makers human? Any simple yes or no answer fails to capture the complexity of what actually happened. Bipedalism had already been around quite a while before the first tools. Hands were changing, but this was a slow process during which the basic toolmaking techniques changed very little. Larger brains were evolving, but there are no known momentous thresholds of rapid expansion. So while it is pretty clear that these changes were related to each other, it is not very clear how.

The history of early stone tools seems to show a slow but steady curve toward greater refinement in techniques. The first tools, the Lomekwian, appear to be the simplest in terms of manual dexterity and cognitive demands. In fact, they suggest that the first stone tool techniques were probably based on nut-cracking, something that chimps do. The use of an anvil fits in here. Lomekwian tools "were made using arm and hand motions that are most similar to the hammer-on-anvil technique used by chimpanzees during nut-cracking," but a bit more cognitive ability is required (Kivell 2015). By contrast, Oldowan tools are made free-hand. One hand holds the target and the other hand is used to hold a stone that strikes the target at just the right angle to chip off flakes.

About a million years after Oldowan stone tools first appear, they became slightly more refined. Some experts refer to these modifications as "developed Oldowan." At about 1.7Myr, stone tools show more flaking and retouching, suggesting a more complicated and intentional process. Larger tools were being made, requiring more intensive work and useful for dismembering even larger animals.

Even so, these early stone tools were relatively easy to make. John Shea comments on his success in teaching college students to make their own tools in roughly the same way. "In teaching flintknapping to college students for more than 15 years, the author has never had a student fail to produce passable replicas of Oldowan cores within the first few minutes of instruction. Other flintknapping instructors report similar results" (Shea 2010). Other experts estimate that the time required to learn basic Oldowan techniques might be more like a few hours. This changes dramatically about 1.7 to 1.8 million years ago when new and more complicated tool-making strategies appear.

Rolling Stones

At this time of change, the simple Oldowan tools are starting to be supplemented by something noticeably more complex, something that experts label "Acheulean" tools. Just as Oldowan tools were named for Olduvai, where they were first found, Acheulean tools were named for the little French village of St. Acheul, were the first tools of this type were found early in the 19th century. Since then, the term "Acheulean" applies no matter where the tools are found, as long as they meet certain other standards.

The most important feature is that Acheulean tools are often flaked on two sides. As the experts say, they are "bifacial." This alone makes them more effective in terms of the sharpness of their edges. They are also large, maybe twice the size of Oldowan tools. And they show more variety in form and probably in their uses than their Oldowan precursors. Acheulean tools include handaxes, picks, and cleavers.

Anyone who has tried to duplicate the making of Acheulean tools quickly learns how difficult it is. Some researchers have tried to define the level of cognitive ability that is required. Of course a high level of hand-eye coordination is required, with fine motor skills. But more than that, these tool-makers seemed to be able to plan their tool-crafting carefully, as if visualizing the design from the outset and imagining it with a kind of 3-D model in their minds. One research team suggests that the tools show that their makers seem "to shape tools purposefully with predetermination or preconception of form." These tools, they suggest, required levels of skill "indicating advanced motor skill and cognition…" The task of making "handaxes, cleavers, and picks suggests imposition of form" (Beyene et al. 2013). Compared to learning how to make Oldowan replicas, learning the style of Acheulian toolmaking is far more difficult. If it takes "a few hours of practice" to learn the basics of Oldowan knapping, "late Acheulean skill may demand hundreds of hours" (Stout 2011). Learning how to make Acheulean handaxes is especially difficult.

We can only imagine how Acheulian techniques were first discovered. We do know roughly when it happened. Recent findings push the date back to more than 1.75 Mya (Lepre et al. 2011; Diez-Martin et al. 2015). Findings at Kokiselei, just west of Lake Turkana in Kenya, date to 1.78 Mya. Meanwhile, hundreds of miles away at the Konso formation in Ethiopia, very similar tools have been found and dated to 1.75 Mya.

It is pretty clear that the Acheulean tool-makers were of the genus *Homo*. But which species? The conventional view was that *H. habilis* made Oldowan tools while *H. erectus* made Acheulean. Now we know that Lomekwian tools predate the emergence of *Homo*. The earliest appearance of *Homo* is 2.75-2.80 Mya, just old enough to support the claim that only *Homo* made Oldowan. But that is far from certain, especially now in light of the discovery of Lomekwian tools.

What about *Homo erectus* and Acheulian tools? Here the evidence seems slightly more strong that *H. erectus* is the inventor and user of Acheulian techniques. Based on what we know now, the timing seems right. *H. erectus* dates from about 1.8 Mya, maybe a little older. The oldest known Acheulian tools were made just after this, at about 1.78 Mya. The fact that they show up in two places almost at once seems to support the idea that *H. erectus*, known for walking long distances, may be the source of the tools in Ethiopia and in Kenya.

Even so, all we have here are coincidences, just enough to trigger a sharp debate among the experts. According to one research team, recent findings suggest that "the earliest known crude Acheulean technology was somewhat widespread by at least ~1.75 Ma. Thus, it now seems that the emergence of the large flake-based Acheulean technology must have broadly coincided with, or closely preceded, the emergence of a H. erectus-like morphology within the early Homo lineage" (Beyene et al. 2013). But according to another team, the connection between Acheulean and *H. erectus* may be overstated. If Acheulean tools arise at the time and in a place where H. *erectus* and *H. habilis* lived, this "strengthens the possibility that more than one tool-making hominin existed at 2.0-1.5 Mya ago" (Lepre et al. 2011).

In fact, just after that report was published, another team presented evidence to suggest that at least one other distinct species of the genus *Homo* lived in the same place at the same time. For some time, experts have debated whether *Homo rudolfensis* is a species or an aberration. With the discovery of new fossils, the evidence is mounting that *H. rudolfensis* is indeed a unique species of *Homo*. The fact that the new fossils were found near Lake Turkana and date to the 1.75-2.00 Mya range makes the question of the Acheulean tool-makers even more complicated (Leakey et al. 2012).

If these scientists are right, their finding complicates even further any attempt to get a clear answer to our question: Do humans make tools or do tools make us human? It could be argued, of course, that it really does not matter. What matters is that in the end, the new technology arises and evolution moves along. Does it really matter who did what? Perhaps only in the sense that these new findings present a complex picture that resists any simple generalization or pattern. It seems that the more we know about our origins, the more complicated things become. And the more complicated our origins become, the more challenging it is to answer simple questions, like what it means to be human. We are not what we once thought we were, a unique, tool-making species evolving over a single track of evolutionary advance.

Sure, tool-use and cognitive abilities are advantageous, but how evolution led to us is not as simple as saying that good tools and big brains win. If it were simple, we might think that *H. erectus* appeared in Africa somewhere around two million years ago and developed Acheulean tools, and with these new tools colonized the world. Two parts of this are true: *H. erectus* appears in Africa about two million years ago and expands fairly quickly throughout Eurasia. But the first humans in Eurasia did not use Acheulean tools. In fact, Acheulean tools do not appear in Eurasia until much later, around 1 Mya (Vallverdu´ et al. 2014). As we are about to see, the same may be true of the technology of the controlled use of fire. *H. erectus* is quick to expand out of Africa, but early technological advances made in Africa seem to stay there for a very long time.

Fire

Learning how to make stone tools was a critical step in our history. So was learning how to control and use fire. Using fire makes it possible to cook food, increasing the nutritional value and expanding the scope of food sources. Fire is used to make some tools because of the way heat modifies the properties of stone or simple chemical compounds.

The Dawn of Technology

Fire also creates a focal point for social life, one that seems to draw us together even today as we sit around a campfire or in front of a fireplace.

If controlled use of fire is so important, when did our ancestors first discover it? "The ability to control fire was a crucial turning point in human evolution, but there is no consensus as to when hominins first developed this ability" (Berna et al. 2012). The problem researchers face here is obvious. Unlike stone tools, fire exists in nature. Finding charcoal or evidence of burning on stones associated with early human remains does not show that the humans were using the fire.

Nevertheless, some claim that the controlled use of fire goes back almost two million years. That is possible but not proven. The best evidence is circumstantial. Beginning around two million years ago, bigger brains and smaller teeth are evident in the fossils. That probably means more nutritional value with less chewing. And that suggests cooking. One research team summarizes the argument this way: "Eating cooked foods made early hominin digestion easier, and the energy formerly spent on digestion was freed up, enabling their energy-expensive brains to grow. Using fire to prepare food made early humans move away from the former feed-as-you-go-and-eat-raw-food strategy and toward the sharing of cooked food around fires, which became attractive locations for increased social interaction between individuals" (Roebroeks and Villa 2011).

Many claims have been made for evidence of controlled use of fire dating back around 1.5 Mya. But other explanations for this evidence are hard to rule out. Were these early humans toasting treats over a campfire or running from a wildfire?

New and more definitive evidence of controlled use of fire now points us back at least one million years. It is quite likely that the use of fire goes back earlier, but for now the million-year discovery date is the earliest one supported by convincing evidence. Working deep inside a cave in the northern part of South Africa, researchers used new techniques of chemical analysis and found what they call "unambiguous evidence in the form of burned bone and ashed plant remains that burning events took place in Wonderwerk Cave during the early Acheulean occupation, approximately 1.0 Ma" (Berna et al. 2012).

But if controlled use of fire goes back at least a million years in Africa and if it is so beneficial to the early humans who used it, we would think that fire played a critical role in making it possible for *H. erectus* to expand out of Africa and into Eurasia. We would expect to find evidence of fire associated with nearly every site of early human occupation, especially in the more northerly and colder regions of Europe or central Asia. When these early humans lived there, the temperatures very likely dropped below freezing.

Despite their living in the cold of the north, there is surprisingly little evidence that the first humans in Europe or Asia used fire. In fact, until 2016, there was no evidence of the controlled use of fire in Eurasia until long after its first use a million years ago in Africa. In a recent review, one research team concluded that evidence for use of fire in Europe is "nonexistent" until somewhere between 300-400,000 years ago. In their judgment, "the European evidence strongly suggests that the habitual and *controlled* use of fire was a late phenomenon" (Roebroeks and Villa 2011). The researchers are not just claiming that we have not found evidence to prove that fire was used earlier. They are suggesting that that evidence seems to prove that it was not used. In 2016, however, scholars published evidence of the use of fire at a site in southeastern Spain. The date at that site is 800,000 years ago (Walker et al. 2016). Whether more evidence of early use will be discovered is an open question. Right now, odd as is seems, evidence of consistent or habitual use is missing.

Accustomed as we are to central heating, we wonder how this is possible. One possibility is a higher rate of metabolism. Even so, this finding is "a surprising one: Where we would expect habitual use of fire…we do not see any clear traces of it" (Roebroeks and Villa 2011). Only much later, less than 400,000 years ago, is there widespread evidence of habitual use of fire by the human occupants of Eurasia. These occupants, of course, were the early Neandertals.

Four Unanswered Questions

The story of human tool-making and fire-use is a surprising and complicated story. Some things seem obvious. Tools and fire are part of a larger process of change in human anatomy and behavior. But just how do these components of human evolution fit together? And how do we explain some surprising features of this story?

We end this chapter by exploring four questions about how our distant ancestors first came to make and use tools, including fire. Today we come to the question of technology through our own recent experience. The rapid pace of technological change in our own time is unprecedented in all of human history. We take it as dogma that newer and better technology is always being developed and that it always replaces what came before. Because of this expectation, we are likely to think that once Acheulean tool-making develops, it will replace Oldowan techniques quickly and permanently.

Yes, there does seem to be a slow curve trending toward tool refinement. The facts, however, are more complicated than that. Not only is the pace of refinement slow, but the spread is erratic. Acheulean techniques appear at least by 1.78 Mya, but Oldowan tools are made for almost another million years, especially in Europe. Humans expand quickly across Africa and Eurasia. Technology moves more slowly. Across the vast geography of our range, some early humans were making Acheulean tools while others were making Oldowan, sometimes almost side-by-side geographically.

Our first question, therefore, has to do with the slow pace of refinement. It is as if each advance—Lomekwian, Oldowan, and Acheulean—achieves a new plateau that remains largely unchanged for hundreds of thousands of years. What are we to make of these long, static plateaus of technological refinement? From 3.3 Mya to about 0.3 Mya, a stretch of three million years, technological refinement reaches only to the level of advanced Acheulean techniques.

Today, of course, we have an unusual view of the pace of technological change. We expect improvements almost monthly. Even if we take a longer view, a million years seems an unbelievably long time for something as sophisticated as Acheulean technology to remain largely static. When we look over the entire scope of our history, we see just how odd we are, we humans who are living today. Technological change for us is so rapid that we cannot keep up. In that sense, our first question is two-fold. How do we explain the remarkable stability of the Oldowan and the Acheulean and, equally, the remarkable dynamism of our era?

The second question is closely related. We cannot help but ask why new techniques took so long to spread. *Homo erectus* expanded out of Africa at the same time that Acheulean techniques were on the rise. Based on what we know today, it seems that it took about 400,000 years for the new techniques to be used in Eurasia. What explains the slowness of this process of transmission? Were the new techniques too complicated for easy use, especially by more isolated groups of humans? Do they require experienced tool-makers as teachers? Were they unneeded when food was abundant, as it may have been where humans were less densely populated? There is some evidence to suggest that the more densely settled the early humans were, the more they used Acheulean techniques. But this pattern of settlement corresponds with geographic distance out of East Africa. The closer we go to East Africa, the human population was more densely settled and more invested in Acheulian techniques. So the answer to the question of the spread of technology might lie in human proximity. Even so, we cannot help but wonder why the first hominins in Eurasia seem not to know about Acheulean techniques. Did they abandon them and forget how to use them? And when the techniques finally appear much later, was it because the Eurasians invented them?

Our third question is how we should explain the similarities in Acheulean techniques from Africa to Eurasia. The techniques being used are not just stable over time. They are also consistent over a vast geographic expanse. From Africa to India to Wales, and for more than a million years, tools are being made in roughly the same way. How do we account for the stability of a technology with a level of sophistication that requires communities of teachers and learners? We do not know if these early humans, probably *H. erectus*, were able to speak much at all. We know they did not speak Acheulian. There was no Acheulean culture or civilization. And yet there seems to be a virtually unbroken web of transmission.

Of course the techniques could have arisen more than once, and they probably did. But even so, overwhelmingly they are transmitted, not invented. The human capacity for imitation, communication, teaching, and learning all seem to be required by the very existence of this technology. We have no access to such intangible features of early human life. But from the tools, and even more by the fact that they are found most where human settlement is most dense, we wonder about the connection between our anatomy, our technology, and our increasing reliance on community.

This leads directly to our fourth question. How should we understand the relationship between tools, behavior, and anatomy? Everything about us was changing. How do the changes fit together? Once it was thought that becoming bipedal was the critical factor that caused our ancestors to use tools. The thinking was that if their hands are free, our predecessors will pick up stones to smash things, then modify the stones to cut things. As plausible as this seems, we know that what actually happened is not this simple.

For one thing, bipedality goes back to the beginning of the hominin lineage. Ardi walked upright, but there is no evidence that she made tools. Very likely she would have cracked nuts with stone, much like chimps. But walking upright is not something that happens suddenly in our past, and it does not trigger a sudden change to toolmaking. These changes seem to be more gradual. Ardi's hands lacked the dexterity that comes with changes in bone structure evident in *H. erectus*. But as far back as *Australopithecus* and even to *Ardipithecus*, hominin hands were able to use stones as tools. "Together recent evidence suggests that pre-Homo hominins were more dextrous than has been traditionally assumed, that tool-related behaviours have played a chronologically deeper and more prominent role in our evolutionary history than previously considered, and that the hands of these early hominins were capable of combining the functional requirements of both arboreal locomotion and enhanced manipulation" (Kivell 2015). No one is claiming here that *Australopithecus* had the advantage of the precision grip or overall manual dexterity of *Homo erectus*, just that the changes in that direction go way back and take a long time to develop fully.

As our predecessors spent less time in the trees, their hands became even more able to hold stones with enough dexterity to strike one against another with precision. Changes in hand structure, brain capacity, arched feet, diet, and pair bonding were all taking place, not quickly of course, but in ways that played off each other, reinforcing and making each other possible. In time, these changes led to even greater advances in technology and to the first expressions of symbolic culture, topics that I will explore in Chapter 8. First, however, we need to ask about the emergence of the genus *Homo*. Who were these first members of our genus?

5

Genesis and Exodus: The Genus *Homo* Travels the World

The biblical books of Genesis and Exodus tell the story of Adam and Eve, the first humans. According to Genesis 1, God creates the world by separating light from dark and the seas from dry land. Then God commands the seas and skies to bring forth fish and birds. The earth brings forth the land animals. Finally, God creates human beings, and from the start they stand out from everything else. Of all the creatures, God declares that only the human pair are created in the image of God, made to reflect God's own glory. As the story unfolds, the descendants of Adam and Eve become the diverse groups spread across the ancient world.

In this chapter, we tell a different kind of story about human genesis and exodus. Like Genesis, our focus here is on the human genus. Like Exodus, we tell the story of the migrations of the earliest humans. And in the end, our purpose in telling the story is the same. Like the ancient Biblical writers, we are asking profoundly basic questions about our humanity. Who are we? As a genus, where did we come from? As a species, where are we going? Of all the creatures, do we have a special relationship with God?

At this point in the telling of our story, however, we are drawing on scientific findings. Chapter 5 is an attempt to describe the origins and the spread of the genus *Homo* based on the insights of the latest science. Our objective here is to be as clear and as accurate as possible. But more than that, our goal is to do more than summarize the latest research. We want to work our way from the best science to the strongest, clearest, most honest theology. How do we understand *this humanity*, this evolved human genus, in its relationship with God? As we work our way toward the final chapters of this book, our goal is to provide the most compelling contemporary theological perspective on humanity possible.

Why *science and theology*? Science is indispensable. It is our best pathway to understand what actually happened. But science cannot answer all our questions. Science can point to how unusual we are among species. Who were the first humans? Where and when did the genus *Homo* first appear? How did they spread throughout the world? Science helps us answer these questions. But in doing so, it also hints at ways in which our definition and identity are hidden from us. The more we probe the sciences of human origins, the more clearly we see that our identity is complex and enigmatic. For that very reason, it is good for theology to ask what a contemporary scientific retelling of genesis and exodus looks like. When it comes to understanding human beings, science helps make good theology possible.

In its shortest form, the scientific rendition of Genesis and Exodus goes like this. More than two million years ago, the first humans appear. Compared to us, they seem hardly human at all. They were shorter than us and more robust. Their brains were scarcely larger than those of today's apes. They had short childhoods followed by a brief and early adolescence. They faced death shortly after the birth of the next generation, if they were lucky enough to make it that long. They used simple stone tools, much like the ones their pre-*Homo* ancestors had made for almost half a million years. In time they spread throughout the Old World. From their fossils we can see that over time, their bodies and brains and behavior were slowly changing. Within a million years, they were becoming taller and their brains started to become larger until they more than doubled in size. They began using more advanced tools and controlling the use of fire. Within another million-plus years, they gave rise to us.

That is the quick summary of the new human genesis and exodus as told by today's sciences. In almost every detail, it is different from the Bible. According to the biblical Genesis, God creates the first humans in a flash of divine action. There is nothing gradual about the origins of the biblical Adam and Eve. They have no biological ancestors. Their appearance is sudden, and from the first moment of their existence they are separate from all other creatures, decidedly more intelligent and complex.

According to the scientific version of genesis, on the other hand, the arrival of the first humans is not marked by sudden bursts or clear boundaries. There are no unambiguous markers, no defining lines drawn in the evolutionary sands, no visible historic thresholds between human and nonhuman. As hard as we might search for the boundaries, we cannot find them. Instead, when we look back at the fossils, the teeth and the skulls and the other fragments that bear witness to our origins, we can discern no bright lines that distinguish humans from nonhumans or pre-humans. There are no faint lines, no lines at all, in fact, that mark the advent of the human, unambiguously distinguishing our kind from other kinds or announcing our arrival.

The Human Question Mark

For a little more than a century now, scientists have dug through caves and stream beds looking for bones. When blessed with the right combination of good guesses and good luck, they uncover physical remnants of human ancestry, odd bits and pieces of our predecessors. Thousands of samples have been discovered and labeled.

It would be nice if the labels or identifying tags came already attached to the fossils at the time of their discovery. Better yet, it would be really helpful if all the fragments of our past came with pre-attached species names. The best we have is expert analysis of their physical features and their context. Fossils provide facts about the past, but facts require interpretation. Only through disciplined observation and informed debate do the stones tell their stories. The direct fossil evidence, of course, constrains all our interpretations. But when it comes to classifying and categorizing and constructing the story of what might have happened, we have to move beyond direct knowledge to expert opinion.

Today, expert opinion about human origins has converged around this general picture: More than two million years ago, while the genus *Australopithecus* was gradually disappearing, the first members of the genus *Homo* began to appear. Slowly over time, something new was emerging. It was something that changed our planet forever, something that in the end deserved to be called human. But humans did not appear as human in a flash. It is not as though someone woke up one day and was suddenly human. There is nothing abrupt or discontinuous about becoming human. And for that very reason, our origins are puzzling, even tantalizing. When we seek to know our origins through science, the main reward for our efforts is a more detailed understanding of a complex and confusing process. The findings themselves almost seem to mock our efforts to come to any quick clarity about our distant past.

Maybe that is just the way it should be. After all, in almost every respect, we humans are hard to figure out. Morally and spiritually and philosophically, we are enigmatic, mysterious, and inaccessible to ourselves. What is human nature? What is a human mind or person or self? We can point to these things, but we cannot really define them. And if we cannot understand ourselves, why would we think we can understand our origins? Is it not entirely fitting that a clear view of our origins, like nearly everything else about us, is almost entirely hidden from our view?

According to Donald Johanson, a leading researcher best known for the discovery of Lucy, "The transition to *Homo* continues to be almost totally confusing" (Balter 2010). Another respected expert, Bernard Wood, begins a technical scientific article by frankly admitting that "the origins of our own genus remains frustratingly unclear" (Wood 2011).

One reason for the confusion is that we simply do not have enough information. Where we would like complete fossilized skeletons, we have a few bones here or a few teeth there. Another reason for the confusion is that new discoveries and new technologies add new information, sometimes surprising information. An even bigger reason for the uncertainty is that it is extremely difficult to put all the known pieces together into a coherent and comprehensive story of what happened. Different experts have different views, and while their debates help advance the field, it is often not clear who, for now, is right.

But perhaps the biggest reason of all for confusion is that there is no simple story to be discovered. The more we learn, the more it seems that there is no clear or straight-line story to be told. It is not the case that one species simply evolves into another. If that were true, then the experts on human origins could arrange the samples in a straight line from least to most evolved. But evolution does not work that way. It is a complex process of branching and rejoining. Populations diverge. They contract and expand. They adjust by behavior or adapt by genetic modification to climate fluctuations or regional circumstances. Then they reconnect, compete, probably fight, probably interbreed, only to diverge again in this dynamic process that we sometimes blithely call "evolution." And when something or someone new turns up, even something sort of human, all we can say is "they appeared." Whatever evolution actually means, it does not mean a straight line or linear process of development from simple to complex.

The problem here is only more acutely felt because we human beings seem to like simple stories with straight story lines. We like to think that one human lineage gradually evolved from the simplest past to the complex present. "It is somehow inherently appealing to us to believe that uncovering the story of human evolution should involve projecting this one species back into the past: to think that humanity has, like the hero of some ancient epic poem, struggled single-mindedly from primitiveness to its present peak of perfection" (Tattersall 2012, 87). Evolution is more complex than that.

It is tempting to think that if we were just better at science, if we just knew more, the complexity would disappear. Some might even think that if the story science tells about our origins is too complex, the science must be wrong. But in fact the cause of the complexity lies in nature, not in science. Nature itself is complex. Figuring out what happened two million years ago is a challenge even if it were simple. But the more we know, the more we see the inherent complexity of our past. Discovering this complexity does not disprove science. If anything, it is evidence that we are on the right track.

So, yes, even the experts who disagree on just about everything tend to agree that today is a confusing time in the field of human origins research. But it is also a wonderfully exciting time. Never before have we known so much. More than that, never before have we been learning so rapidly. There has never been a more productive time to be learning about human origins. Intense debate among the experts drives new research. Unexpected discoveries come flying at us before previous findings are fully comprehended. All the while we are forced us to consider and reconsider older perspectives.

We wish we knew more, but we already have enough information to answer the basic questions about our past. In light of the latest discoveries, what can we say about the origins of the genus *Homo*? Who were the earliest members of our genus? When and where did they first appear? What might life have been like for them, and what became of them? According to the latest research, here is what we can say.

The First Humans

For several decades, the most commonly accepted view is that the genus *Australopithecus* is the direct ancestor of the genus *Homo*. Sometime before 2 Mya, *Australopithecus* gave rise to *Homo*, and humanity appeared. In the broadest terms, that version of events is generally correct. But the more we dig into the details, the less adequate or accurate the simple version seems to be.

For one thing, it turns out that during the very time when the first humans appeared, there were several forms of *Australopiths* just about to exit the stage. It also seems that there were several forms or species of *Homo* emerging from the shadows in the wings. So the obvious question is which species of *Australopithecus* gave rise to the first species of *Homo*. The question, however, is misleading. There is no last *Australopith* or first human. The more fossils we have, the less they fit into a coherent and neatly sequential family tree. In fact, a new picture is emerging. Several distinct forms of *Australopithecus* are changing independently into some of the more "derived" or human-like forms that follow. It has been suggested that *Homo* is the result of interbreeding, emerging in a new form as a result of the coming together of traits and features that first appear in independent lineages.

A big problem here is that we really do not have nearly as many fossils as we would like from the period of 2-3 Mya. This is the critical epoch of human emergence, and we have next to nothing in the way of direct evidence. We do know something about the climate fluctuations of this period and about the other plants and animals that lived then. We also know about the hominins that came before and those that came after. But our direct knowledge of this key transition is limited to what can be interpreted from isolated fragments.

The dramatic announcement of one such fragment occurred in 2015. A team led by Brian Villmoare reported on a partial jawbone that dates from 2.75-2.8 Mya. Found in 2013 at Ledi-Geraru, Ethiopia, the fossil known as LD 350-1 is half a lower jaw with some well-preserved teeth. Experts tend to agree that in some ways, it resembles later fossils that are clearly *Homo*. On that basis, the authors claim that LD 350-1 "establishes the presence of *Homo* at 2.80-2.75 Ma" (Villmoare et al. 2015). Not surprisingly, some wonder how this isolated fragment can establish anything as big as the dawn of humanity (Hawks et al. 2015). Is LD 350-1 best classified with *Homo* or with *Australopithecus?*

Defending their view that LD 350-1 is best seen as early *Homo*, Villmoare's group explains that their find resembles "younger east African *Homo* specimens more than it does geologically contemporaneous or older *Australopithecus.*" This means that the genus *Homo* first appears by 2.8 Mya, at least in a proto-*Homo* form they call an ancestral population. In their view, LD 350-1 is *Homo* even if it is not clear which species. So they conclude with this claim: "For the present, pending further discoveries, we assign LD 350-1 to *Homo* species indeterminate" (Villmoare et al. 2015).

Does an isolated, partial jaw really provide enough information to support the claim that the genus *Homo* goes all the way back to 2.8 Mya? The claim would seem more plausible if we had other fossils almost as old that clearly belonged to *Homo*. Right now, the one that comes closest chronologically dates from 2.3 Mya, almost half a million years after LD 350-1. The later fossil is also part of a jaw, this time the upper part or maxilla, along with a few teeth. It was found in Hadar, Ethiopia. Like LD 350-1, the Hadar fossil is fragmentary and does not tell us very much. It does seem to fit with some other bones that date from a little later, and so many experts agree that the Hadar find belong to the genus *Homo*.

If Hadar is *Homo*, why not LD 350-1? At the moment, there is just not enough evidence to settle the debate. The advice from Hawks and his collaborators is to leave the question open. They "urge caution when assessing the taxonomic affinities of such isolated remains, because at present we cannot be certain what the rest of the dentition, skull, or skeleton of LD 350-1 might have looked like" (Hawks et al. 2015).

It is only around 2 Mya that things start to become clear, and then only partly. The problem in the 2-3 Mya timespan is the lack of fossils. The problem after 2 Mya, however, is how to fit all the fossils together into a coherent pattern. Researchers can see that there is a great deal of diversity among the hominins living then. Some argue that we should expand the scope of known species to include most or all the diversity in one or two species. Others believe that it is best to think that several species of *Homo* existed at once.

Homo habilis, of course, is the most familiar form of early *Homo*. But some fossils classified as *habilis* are hardly any different from *Australopithecus*. And as a collection, the fossils show that there was a lot of diversity in the period between 1.8-2.0 Mya. In light of this, experts are debating whether *Homo habilis* really *Homo* at all. It is pretty clear that *habilis* made tools. But toolmaking is no longer a trait limited to the genus *Homo*. Perhaps it is better to see *habilis* as transitional to *Homo*, more a ramp than a new level, a passage from the past rather than the dawn of the present. In that case, the older jaw fragments that date from 2.75 and 2.3 Mya may fall within a broad notion of *H. habilis*.

One of the better examples of *habilis* is known in the archeology records as KNM ER 1813, found at Koobi Fora, Kenya. It is relatively complete cranium but nothing more, and it is dated to about 1.9 Mya (M. Leaky et al. 2012). For *habilis*, it seems to have an unusually small brain. The size is about 510 cc. That is approximately the same size as the more recently discovered *A. sediba*. Is the brain of *habilis* large enough to be considered human? How large must a brain be in order to be human?

In addition to asking whether *habilis* really belongs in the genus *Homo*, experts also debate how many other species of *Homo* may have lived during the transitional period between 1.8 and 2.0 Mya. New findings are being reported and previous findings are being reassessed using new techniques. The result is an unexpected amount of diversity among specimens that seem to belong clearly in the genus *Homo*. For example, worldwide attention was drawn in 2015 to the announcement of a new species, *Homo naledi*. A team led by Lee Berger discovered a trove of new fossil remains deep in a cave in northern South Africa (Berger et al. 2015). Bones from at least fifteen different individuals were found. While the *naledi* group shows relatively little diversity, fitting *H. naledi* into the wider genus is not easy.

The single biggest problem with the *H. naledi* samples is assigning a date. They were located in a setting that makes the dating process especially hard, and so far the experts studying these fossils have not tried to say when *naledi* lived. From their physical traits, it looks like *naledi* lived about 2.0 Mya or just a bit later. But experts cannot rule out the possibility that these *naledi* samples were part of a late-surviving group. And while they seem to fit at about 2.0 Mya, they do not match anything else from that period. They seem instead to have features or traits that are a mosaic, a bit of this and that, in some ways like *Australopithecus* and in other ways like *Homo*. All in all, however, Berger and his colleagues insist that *naledi* fits better with *Homo* than with *Australopithecus*. Berger suggests that *naledi* is most like the very earliest forms of *Homo*, "rooted within the initial origin and diversification of our genus" (Berger et al. 2015).

Not everyone agrees with Berger's classification of *naledi*. But there is growing agreement that in the 1.8-2.0 Mya period, *Homo* is diverse in form. Commenting on the discovery of *H. naledi*, Chris Stringer had this to say: "Even without date information, the mosaic nature of the *H. naledi* skeletons provides yet another indication that the genus *Homo* had complex origins. The individual mix of primitive and derived characteristics in different fossils perhaps even indicates that the genus *Homo* might be 'polyphyletic': in other words, some members of the genus might have originated independently in different regions of Africa" (Stringer 2015). The diversity in early *Homo* seems to suggest that the traits that define our genus did not evolve in one neat lineage or one distinct community, much less in one clean burst of evolutionary advance. It seems instead that our defining traits arose in diverse, geographically dispersed communities. According to Berger's group, the "accumulating evidence has changed our perspective on the rise of our genus" (Berger et al. 2015).

This is true not just for anatomy. It is very likely also true for hominin behavior. According to Berger, various features "including increased brain size, tool manipulation, increased body size, smaller dentition, and greater commitment to terrestrial long-distance walking or running" were once described as a "a single adaptive package" that arose as a whole (Berger et al. 2015). Now it seems that each part arose independently and then converged in various ways. This is what explains the variation seen in early *Homo*.

The significance of *Homo naledi* is not that it fits our preconceived ideas, but that it does not fit. The Berger team sees *naledi* as an unexpected combination or mosaic of traits. "This species combines a humanlike body size and stature with an australopith-sized brain; features of the shoulder and hand apparently well-suited for climbing with humanlike hand and wrist adaptations for manipulation; australopith-like hip mechanics with humanlike terrestrial adaptations of the foot and lower limb; small dentition with primitive dental proportions."

Then, as if to cast more doubt on the findings from 2.75 and 2.3 Mya, the Berger group argues that *naledi* seems to show that other complete specimens might also show an unexpected mosaic or combination of trait. If only a jaw is found, and it looks like *Homo*, is it really safe to classify it based only on knowledge of one part of the anatomy? According the Berger, *H. naledi* seems to suggest that "in light of this evidence from complete [*naledi*] skeletal samples, we must abandon the expectation that any small fragment of the anatomy can provide singular insight about the evolutionary relationships of fossil hominins" (Berger et al. 2015).

At this critically important period right around 2 Mya, the earliest examples of our genus *Homo* are diverse and hard to classify. Things get a little better when *Homo erectus* arrives at about 1.8 Mya, but only slightly so. In particular, the relationship between *H. habilis* and *H. erectus* is enigmatic. According to Bernard Wood, *H. habilis* "is too unlike *H. erectus* to be its immediate ancestor, so a simple, linear model explaining this stage of human evolution is looking less and less likely" (Wood 2014). The pathway between *habilis* and *erectus* is no straight line but more like rabbit trails through a thicket. At least several distinct communities if not distinct species were involved. Nor did *H. habilis* simply vanish when the first members of *H. erectus* appeared. Some fossils dating from 1.4 Mya seem to fit best with *habilis*. And if it is the human "adaptive package" that is critically important in defining us, any compelling answer to the question of how it came into existence eludes us.

But a new combination of traits seems clearly in place by about 1.8 Mya, when *Homo erectus* appeared on the scene. With *Homo erectus*, something new arrived, something more like us but something that raises even more questions.

Walking Tall, But Which Way?

In Africa, the earliest confirmed date for *H. erectus* is 1.78 Mya. This is based on analysis of a relatively complete cranium found at the Koobi Fora site in Kenya. While this skull is dated fairly precisely to 1.78 Mya, many experts believe that *H. erectus* appeared in Africa well before this date, at least by 1.9 Mya and perhaps as far back as 2 million years. Other fragments from this earlier period have been found in Kenya and Tanzania, but nothing is so complete or definitive as the Koobi Fora cranium (Leakey et al. 2012).

A big reason for thinking that *H. erectus* must have appeared before 1.78 Mya is that by that time, they seem to have traveled a long way. Remains found much further south at Swartkrans, South Africa, appear to belong to the same species and date to almost exactly the same time. Even most striking is recent evidence that they lived in modern Georgia, east of the Black Sea, as early as 1.85 Mya (Ferring et al. 2011). Fossils very much like *H. erectus* and dating to 1.77 Mya have been found (Lordkipanidze et al. 2013). But just below these fossils is a slightly older layer dating to 1.85 Mya, not containing any fossils (at least so far) but containing simple stone tools. This is pretty strong evidence that some form of human life, probably some version of *Homo erectus*, was living there just before the first definitive evidence of their existence in Africa.

If *H. erectus* did originate in Africa, it is more than a little odd that oldest evidence of their existence is not found in Africa but in southwest Asia. For some experts, this suggests the unconventional possibility that *H. erectus* evolved in Asia and then spread to Africa. This is not the majority view and may never be widely accepted, but it does provide one solution to the problem. The other solution is that fossil hunters simply have not yet found the earliest possible evidence of *erectus* in Africa, or that any such evidence vanished long ago.

But it is not out of the question that these early humans evolved in Eurasia and migrated *into* rather than *out of* Africa (Wood 2011). In that case, what may have happened is that an earlier form of humanity—*Homo habilis* or something like that—first appeared in Africa and then migrated to Asia. Once there, it continued to evolve until it became a somewhat early form of *H. erectus*, resembling what was found in Georgia and quickly spreading to Africa.

For now, we simply do not know where *H. erectus* came from. The more we learn about their existence in the period prior to 1.5 Mya, the more puzzling they seem. They are more diverse than we once thought, so much so that experts debate whether *H. erectus* is one species over its vast territory and lengthy species lifespan, or whether it is best thought of as several species, such as the Asian *H. erectus* and an African counterpart, *H. ergaster*. These are debates not over facts or even interpretation of facts. Everyone agrees that there is a high degree of variability here. It is mostly a debate over abstract concepts like species and how much variation can be contained within a single species (Hublin 2014).

Before very long, *H. erectus* appears in eastern and southern Asia, having made the long trek to modern China and even to the islands of Indonesia. Their first appearance in Europe is a bit later (Muttoni et al. 2010). Somewhere between 1.1 and 1.2 Mya, we know they were living in northern Spain. What is more amazing is that as a species, *Homo erectus* survived for more than a million years, probably 1.5 million years or more, depending on how certain remains are classified. That is a species survival span about ten times longer than our own *Homo sapiens* survival so far.

What Were They Like?

We may not know where they came from. And as we will see, we do not have a very clear idea of what happened to them or how it happened that they set the stage for the next phases of human evolution. But from the many bones and teeth and tools they left behind, we can tell quite a bit about them.

The standard view of *H. erectus* is that here at last is a form of human life that really seems to resemble us. They were often about five feet tall, sometimes perhaps as much as six feet. Some believe they grew up somewhat as we do today, with a drawn-out infancy and lengthy childhood capped off by a late adolescent growth spurt, or at least that they were tending toward changes in that direction. Early in the long span of their evolution, their brains tended to be in the range of 600-800 cc, roughly half the size of human brains today. At the lower end, that is about the same as *H. habilis*, prompting some to ask why the humanness of *habilis* is questioned while the humanness of *erectus* is secure. In fact, the new findings from Georgia seem to obscure any sharp line between *H. habilis* and *H. erectus*.

Toward the end of the evolution of *H. erectus* as a species, their brains are reaching the 1000 cc size. That is still notably smaller than ours, but getting close. For members of one species to vary from about 600 to 1000 cc in brain size, however, is beginning to stretch our concept of a species. Should they all be classified as one species? But what else might we expect of a species over time if not some significant change?

Providing food for the development and the nourishment of such large brains and bodies required behavioral and technological changes. Just how all the components fit together remains an open question, but larger brains and a longer period of infancy are linked with better techniques for getting richer food from more varied sources. Pregnant and nursing mothers required protection and access to rich and consistent sources of food. At the same time, longer legs and a more graceful gait plus the ability to cool the body by perspiration made it possible to range over greater territories in pursuit of food. By about 1.78 Mya, at least some *H. erectus* communities were making the transition from Oldowan to Acheulian tools, which literally opened up new sources of food. Teeth are becoming smaller, especially the molars. The digestive tract is smaller and uses less energy than before. As the chewing and digesting of food take less energy, the expanding brain compensates by needing more calories than ever before.

One study summarizes these changes this way: "This dietary shift to more energy- and nutrient-dense foods would potentially have allowed for an increase in brain size by removing constraints on brain growth; in addition, this dietary change may be selected for increased brain size and cognitive capacity related to increased foraging, extraction, and processing abilities associate with higher-quality diets. The reliance on high-quality foods may have also selected for cooperative breeding. . .to support the growth and high maintenance costs of large brains (Antón and Snodgrass 2012).

A key transition during the long reign of *H. erectus* has to do with their ability to store fat. Human beings today are remarkably adept at storing fat. Most of us would rather not have this capacity, but it is key to the survival of any large-brain, energy-burning animal. High energy food is not always available, and there is growing evidence that *H. erectus* solved the problem of high demand in the face of variable supply by evolving the capacity to turn calories into stored fat to be burned off when needed.

And then somewhere during all of these changes, probably at least by 1 Mya, the controlled use of fire made cooking possible. This aided digestion even more and opened up new sources of food while also creating a focal point for communal life, perhaps with profound consequences beyond the value of a warm meal.

Skeletons are becoming somewhat less robust and more fine-boned or "gracile," more like us but still pretty robust by comparison. All these changes occurred slowly over time. But the main point to emphasize is the *H. erectus* was a wide-ranging species in every respect. They were widely distributed geographically, existing for more than a million years, and widely varied in physical and behavioral traits. More than that, they seem to have possessed a growing flexibility. Not only did they differ from each other, but they seem to have become more able to adapt or modify themselves when faced with different circumstances. In other words, their prodigious ability to travel seems to have been matched by a remarkable ability to adjust, adapt, and survive in a new or changing environment.

The Importance of Being Adaptable

While we do not know for sure whether they evolved in Asia and migrated south or in Africa and traveled north, we do know that they spread quickly and lived in highly varied climates. They traveled to places that were different from their starting point. On top of that, climate changes came to them. Their climate was constantly changing, and creatures that could not adapt quickly would die. No doubt early death was common among them, but as a community they were adaptable enough to survive. Faced with "habitat unpredictability," early *Homo* seem to have possessed what we call *behavioral and developmental plasticity*, and this increases over time (Anton et al. 2014). While all populations respond to evolutionary selective pressures and evolve genetically, some are able to adapt in behavior and even in physical traits without waiting for evolution. Humans today are able to modify their behavior and to some extent their physical development in response to environmental changes. The evidence suggests that *H. erectus* was well on the way to acquiring this sort of inherent adaptability, making it possible for them to live in widely different environments. This in turn would have exposed them to different selective pressures, affecting their future evolution.

Probably more than anything else, it was their relatively large brains that helped *Homo erectus* be flexible and adaptive when faced with changing environments. But the relationship was probably two-way. The evolution of bigger brains with more flexible behavior made survival in wider-ranging environments possible, and exposure to wider-ranging environments moves evolution a bit more quickly. Not surprisingly, one of the newer questions in human origins research is to ask what can be known about their environmental challenges. New studies in climate change can be correlated with human evolutionary timelines (Anton et al. 2014). Climates change over time, most dramatically in the form of expansion and contractions of snow and ice over most of northern Europe and Asia. In these oscillations, humans were either wiped out or barely survived in the southernmost parts of Europe along the Mediterranean. Small remnant populations are just what evolution needs to produce greater diversity. In partial isolation, small groups sometimes give rise to new subspecies or to distinctive patterns of gene frequency.

Climate oscillations occurred again and again across the extent of the range of *Homo erectus*. No one knows what specific impact this may have had on human evolution. But the challenges of climate change meant that our ancient ancestors were faced again and again with environmental shifts. In the face of change, adaptability is an advantage, and bigger brains increase adaptability. It is clear that over their long span as a species, the *Homo erectus* brains were becoming larger, essentially doubling in size.

With their ability to travel and to adapt to changing environments, we might expect that *Homo erectus* would survive a long time, and they did. They lived at least until about 300,000 years ago, perhaps even later. One intriguing question about *H. erectus* has to do with the Indonesian island of Flores. More than a decade ago, researchers discovered fossil remains of unusually small hominins there in a cave. Were they a dwarfed version of *erectus*? Or contrary to the standard view, did *H. habilis* travel all the way to Indonesia? Or were they more modern humans but with some sort of disease? (Argue et al. 2006).

When the discoveries on Flores were first announced, the media dubbed these early islanders "hobbits," because of their short stature. Their technical name is *Homo floresiensis*. But more than anything, what made their discovery newsworthy was the claim that they lived until somewhere between 11,000 and 13,000 years ago. If so, they would have shared the island with more modern humans for tens of thousands of years.

Now we know that this is not true. In 2016, the original fossils were re-dated to 60,000 and 100,000 years ago (Sutikna et al. 2016). If so, then the last known "hobbit" most likely died about 10,000 years before the first modern humans arrived. However, tools and other artifacts from *H. floresiensis* stretch from 190,000 years ago all the way to 50,000 years ago. This led the researchers who re-dated these findings to end their article with this note: "But whether H. floresiensis survived after this time, or encountered modern humans, Denisovans or other hominin species on Flores or elsewhere, remain open questions that future discoveries may help to answer" (Sutikna et al. 2016).

Also in 2016, we learned that Flores hominins very much like *H. floresiensis* date much further back into the past that anyone imagined. Fossils dating to about 700,000 years ago have now been found on this Indonesian island. They resemble *H. floresiensis*, and in fact are even smaller than the original "hobbits" in size (van den Bergh et al. 2016). Their discovery lends support to the idea that *H. floresiensis* is a form of *H. erectus*. We know that *erectus* lived not too far away on Java from about 1.2 Mya until about 500,000 years ago. Those living on Java were bigger than the "hobbits." They stood about half again as tall, and their brains were nearly twice as large. In 2015, researcher took a closer look at some of these *H. erectus* artifacts from Java. They reported evidence that these early humans were using shells to make tools. One shell in particular caught their eye. On it, they say, is what could be the world's oldest geometric engraving. Straight lines, some resembling the letters M and V, are scratched into the side of the shell. This was wholly unexpected, the researchers report, but they do predict that other discoveries will be made, suggesting a much older history for such engravings (Joordens et al. 2015).

No such engravings have been found on Flores, but scientists have found human artifacts on the island that date to about 1 Mya. For the moment, no one knows for sure how these early humans got to these islands or exactly where they came from. Experts continue to debate who their closest relatives are. Were these creatures small (and more like *H. habilis*) before they arrived, or did they become small once they reached the island?

The hobbit question is not the only possible example of *H. erectus* species longevity. For several years, scientists have been studying remains of the so-called "Red Deer People" in a cave in southwest China. Recent research focused on a leg bone dated to 14,000 years ago. Analysis shows that it did not come from recent *Homo sapiens* but from something much older in our evolutionary past. According to the team that studied the bone, it came from "an individual that probably belonged to an archaic taxon" (Curnoe et al. 2015). What "archaic" form of humanity could that be? It may have been something akin to a Neandertal. Or like *H. floresiensis*, maybe it was another late-surviving remnant population of *H. erectus*.

We may never know the full story of what happened to *Homo erectus*. But perhaps an even bigger question is what came after them. Already by about 700,000 years ago, *H. erectus* is somehow turning into a new form of human with an even bigger brain.

After *Homo erectus:* Bigger Brains and Diverging Forms

Even though *Homo erectus* was still very much present in Africa and Eurasia 700,000 years ago, something new is coming. Bits of evidence suggest that in some places, brain size was expanding from the usual 1000 cc upper end for *H. erectus* to something more like 1250 cc, just below our own brain size. One of the earliest and most important examples is the partial cranium from Bodo in the Middle Awash of Ethiopia dated to 600,000 years ago. Similar bones from the same time have been found at Elandsfontein in South Africa. Over the next half million years, more examples of larger-brained forms of *Homo* appear. Even with their bigger brains, however, these individuals seem on the whole to resemble *H. erectus* more than they resemble us.

We cannot infer very much just from the size of a brain. Its structure or architecture is also critically important, together with the ratio of brain size to body size. Certain cognitive functions like language or symbolic thought depend on specialized structures. For now, at least, we have far too little insight into these areas to speculate very much on whether they engaged in some form of human thought.

In addition, we simply do not know enough yet about these transitional forms of humanity to know whether or not to see them as a distinct species. Years ago, experts spoke of a bridging species called *Homo heidelbergensis*, named for the first example found near Heidelberg, Germany, in 1907. This first example was not a complete skeleton but only a mandible, and it has now been dated to around 610,000 years ago. The teeth are small but the bone is robust. Some experts see the Heidelberg jaw as fitting together reasonably well with discoveries from Bodo, Elandsfontein, and maybe as far away as India and China. These specimens date roughly to 600,000 years ago. In terms of smaller teeth, larger brains, and a less robust frame, they tend to fit together. While some argue that it is reasonable to call them *H. heidelbergensis*, others argue that they should be called *H. rhodesiensis*.

It is not clear how to label these specimens. It is even less clear how to think about the species pathway between *H. erectus* and our own form, *H. sapiens*. One team of researchers summarizes the current state of knowledge (and ignorance) this way: "Although the period bracketed between approximately 900 and 600 ka is very poor of fossil evidence, it seems therefore that something crucial happened at that time, generating a new and more encephalised kind of humanity that spread quite rapidly in Africa and Eurasia. When viewed as a geographically widespread single taxon from which both Neanderthals and modern humans originated...these humans of the Middle Pleistocene should be referred to as *Homo heidelbergensis*..." (Profico et al. 2016). The period is critically important to the story of our origins. Somehow, spread out over Africa and Eurasia, a transitional form called *H. heidelbergensis* gives rise to Neandertals and to us.

When or where did this happen? No one knows. Perhaps a recent finding from Central Asia offers a new clue. In a remote area near the boundary between Russia, China, and Mongolia, Denisova Cave was once occupied by Neandertals and later by more modern humans. Among the many fossilized human bones found in the cave, one tiny fragment stands out. Using the latest techniques, researchers extracted and reconstructed its DNA. What they found was a shock. The DNA was not Neandertal or "modern." When compared with other samples, the finding suggests "the existence of humans that were different from both *Homo neanderthalensis* and *Homo sapiens*, but shared with them a common ancestor between 1.3 Ma and 779 ka" (Profico et al. 2016). This pushes the date of modern/Neandertal divergence back further in time. It suggests that our two lines diverged, not from *heidelbergensis* at around 500-600 thousand years ago, but from something even earlier. This earlier form of humanity is sometimes called *H. antecessor*. If Neandertals emerged in Eurasia and our line emerged in Africa, when did they diverge? What was the last common ancestral community? Is *H. heidelbergensis* or *H. antecessor* best seen as our pivotal ancestral species?

That debate might not be settled anytime soon. Nor are experts likely to agree soon on what happens next. What is the pathway from this earlier form of humanity to the earliest examples of *Homo sapiens*? Where did they come from, and how did they become like us? Is it right to call them "modern"?

6

Are We Modern Yet?

The word "modern" can mean many things. In pop culture, modern refers to the current style. In the field of history, scholars speak of a modern era that stretches back almost to the Renaissance but ends sometime in the last century.

In the field of human origins, "modern" means something quite different. Used here, in fact, it means at least three different things, and these meanings are often not sorted out the way they should be. Most frequently the word is used as a kind of shorthand, a simple way to identify certain human ancestors and not others based on their subtle anatomical differences. Often in technical reports we see the word "modern" used as part of a longer phrase: "anatomically modern humans." Sometimes, however, "modern" refers not to bodies but to behaviors, for instance when researchers refer to "behavioral modernity." At other times, "modern" implies a value judgment that dismisses earlier forms of humans as "archaic" and crowns the most recent as the most advanced. When used this third way, "modern" refers to something much deeper than anatomy and behavior.

The main problem here is that sometimes we slide from one meaning to another without noticing what we are doing. It is worth trying to untangle the meanings of "modern," paying close attention to what we are doing when we use the word to speak of ourselves and our origins. Often we read about ancestors who, at long last, are "anatomically modern" or "behaviorally modern." What do we mean when we say that at some point along the way, our ancestors became modern in anatomy or behavior? Like everything else about human origins, the story about how we become "modern" is more complicated than we might think.

Anatomically Modern

Scientists often use the word "modern" to speak of humans like us. Technical articles frequently speak of "anatomically modern humans" or simply of AMH. Modern humans are usually defined as our direct ancestors who looked pretty much like us. Dressed up like us, they might pass for us, at least enough to fool some people on the street. These modern humans may have appeared as much as 200,000 years ago. But it is important to recognize that the "modern" humans who lived even just 30,000 years ago were on average a bit more robust than us, with a tendency toward more prominent brow ridges. If we go back earlier, the differences are more pronounced. Their bodies and brains were as big as ours and probably had all the same basic features. We should not think, however, that all anatomically modern humans were exactly like us, at least not on average.

The conventional benchmark for modern anatomy is this: An individual specimen from the past is considered to be anatomically modern if its physical traits fit within the range of today's human variation. Humans today vary widely in height and in other physical measurements. That full range of variation provides the parameters. Anything that fits these broad parameters is "modern."

Things get more challenging when the idea of anatomically modern humans is connected to a set of genetic and evolutionary events thought to have given rise to a new and distinctive version of humanity that displaced other forms or species. This new version of humanity arose in Africa and spread to the rest of the world, and so the phrase "Out of Africa" is often used to describe these events. The challenge here is to avoid misunderstanding how "Out of Africa" and notions like "anatomically modern" fit together with the latest research. The best way to do that is to look back over the past century in order to contrast "Out of Africa" with other views.

Not many years ago, experts in human origins tended to think that after *H. erectus* migrated to all parts of the Old World, evolution continued in each part. Developments in Africa, Asia, and Europe continued on the same pace, but in partial isolation from each other. Most thought that there was frequent contact and interbreeding between partially separated groups. Contact was always frequent enough so that the global population remained one human species. There were local or regional differences in dispersed groups, but there were largely superficial in nature. This view is often called "multiregionalism."

Other experts once thought the human evolutionary isolation was virtually complete. On the African, Asian, and European continents, ancient humans evolved independently. The result was three different human "races." This view is called "polygenism." For polygenism, the differences between the "races" are more than superficial. Anyone holding this view is likely to see humanity as defined by racial difference. Polygenism has not been held by experts since the early 1960s, and then by only a few. On the racist fringe, however, polygenism does not go away because it seems to offer scientific support for racism. Today there is full agreement by all the experts that polygenism is clearly wrong because it is inconsistent with the fossil and genetic data. And of all the wrong theories, "the most pernicious is polygenism, the theory that human races have long, separate, and isolated histories" (Hawks and Wolpoff 2003).

The tide of expert opinion turned against polygenism in the period between 1930 and 1960. It turned against multiregionalism in the 1990s. All human beings alive today, it is now believed, are direct and recent descendants of a distinct group that arose in East Africa about 150,000 to 200,000 years ago. They spread quickly throughout the African continent and then on to Asia and Europe. Not surprisingly, this is called the "Out of Africa" theory. In its most strict or pure form, it held that the new humans completely replaced all other forms of humanity already living across Africa or in Eurasia.

In the past decade, however, a modified version of the "Out of Africa" theory has emerged. The anatomically modern humans coming out of East Africa met other, not-quite-so-modern people. The result of these encounters, at least to some extent, was interbreeding. Thanks to advances in genetics, researchers have been able to reconstruct the DNA of Neandertals and other earlier forms of humanity based on tiny samples from bones that are tens of thousands of years old. Were Neandertals or the other groups anatomically modern? Not if we apply the common rule. Their anatomy does not fit within the range of today's humans. Nor were they direct descendants of the new or "modern" form of humanity. What happened when modern humans met Neandertals or other groups? We will probably never know whether their encounters were mostly conflicted or cooperative. We do know that there was sexual activity that resulted in "hybrid" or cross-bred individuals, some of them healthy enough to live and to procreate successfully.

The discovery in the past decade of modern/nonmodern interbreeding has forced experts to rethink the "Out of Africa" theory. In its strict form, the theory held that modern humans completely replaced everyone else. Now we know that this is wrong (Alves et al. 2012). Today's modified Out of Africa theory takes interbreeding into account. The theory still seems broadly true in its central claim. A new form of humanity did get its key start in East Africa between 150,000 and 200,000 years ago. But even there, the evolutionary story is likely far more complex than the classic Out of Africa theory once thought. And that complexity is further complicated by successful interbreeding, which probably happened not just once or twice but often and everywhere (Groucutt et al. 2015).

These findings lead to a fundamentally new view of the humanity that first emerges in East Africa. "Modern humans" are not some new and isolated species. They are not genetically isolated, unique in their anatomy, or definitively *Homo sapiens* from the start. In fact, the new form of humanity may very well have come into existence, not by mutation, but by interbreeding or hybridization. And it may have continued to change by interbreeding as it expanded across Africa and to Eurasia.

Of all the groups that interbred with these new humans, we know the most about Neandertals. We will take a closer look at them in Chapter 7. Here in Chapter 6, our focus is on how the new form of humanity emerged in Africa and made its way into Asia, Europe, and eventually to the Americas. Experts in the study of human origins agree that something new is appearing in Africa somewhere between 100,000 and 200,000 years ago. By all accounts, the appearance of this new human variant is a decisive event in the long story of our human origins. But what do we really know about them? Given their significance, it is perplexing that we do not have more information about who they were.

Some of the oldest known fossils that can be classified as *Homo sapiens* come from sites along the Omo River in southwestern Ethiopia (Aubert et al. 2011). The fossils are dated to 195,000 years ago. Two skulls were found on the site. One seems to fit well with *H. sapiens* while the other seems different and somewhat questionable in terms of fit. The most obvious explanation is that at the dawn of *H. sapiens*, there was a fair amount of diversity even in what was probably a small population. Taken together, the two skulls seem to reflect a transition rather than a new form of humanity.

Another important set of early fossils has been found at Herto in the Middle Awash of Ethiopia. They date to between 154,000 and 160,000 years ago. They offer more convincing evidence of what are usually described as the new "modern" humans. These remains were found in 1997, when researchers found the skull bones of three individuals, including two adults and a child estimated to be about six or seven years old. These individuals lived in Ethiopia at the same time the Neandertals lived in Europe and western Asia, but the two populations are noticeably different. Compared to Neandertals, these Herto specimens were less robust. Simply put, they looked somewhat more like us than any other human remains from the past. At the same time, there are also marked differences between these Herto fossils and human beings living today. The Herto individuals may have been less robust than their contemporaries, such as the Neandertals, but they were still more robust than we are (White et al. 2003).

Dating from just a few tens of thousands of years later, findings in Ethiopia and Morocco provide more hints into what seems to be emerging. First in East Africa but now also in the continent's far northwest corner, a less robust, somewhat taller form of humanity is showing up. Only a little later, similar remains appear for the first time out of Africa. In modern Israel at two cave complexes known as Skhul and Qafzeh, the remains of numerous individuals have been found. These date from between 80,000 to 120,000 years ago.

If they were standing next to us, they would seem relatively more robust and tend to have what are commonly regarded as "archaic" features. But they are clearly similar to the new version of "modern" humans. When we say this, we have to recognize that "modern" is a relative or comparative notion. Compared to their contemporaries, they were somewhat modern. Compared to us, they were less modern. Whoever these new humans were and regardless of how much they stood out from their contemporaries, it is certain that all human beings alive today are their descendants and share their modernity. Because this is true, it is easy to see how in the discussion of human origins, the word "modern" also means living humans or *Homo sapiens*.

Why did this new form of humanity appear when it did? Several factors may have been in play. The environment may have been changing, requiring rapid changes in order to survive. New food sources, such as protein-rich shellfish, may have become available. New tools may have been developed, offering access to other food sources. And of course, genetic mutations and changes in gene expression may have occurred, changing the way human brains grow and develop. Or it may have been that interbreeding between two distinct lineages gave rise to something new. Acting alone, none of these factors is enough to explain what seems to have happened. But it is clear that changes like these occurred, leading to a community of new humans that are ancestors of us all. And it also seems certain that this community of common ancestry diverged, changing and adapting to become slightly different as groups went in different directions.

From roughly 1990 to 2008, many experts saw the new humans as a distinct species that completely replaced everyone else. This notion was advanced in the late 1980s and was associated with the claim for a "mitochondrial Eve," a hypothetic mother of all living humans (Stoneking and Cann 1987). This "mitochondrial Eve" was connected to the fossil evidence for the Out of Africa theory. The result was the exaggerated claim for "an *African origin* for a *recently evolved modern human species*" that spread throughout the world and completely replaced all existing populations without interbreeding or admixture (Wolpoff and Caspari 2011; italics in original). Today, of course, we have good reasons to believe this is not true.

These new humans were not a new species, but they were something relatively distinct and different. They entered new lands, but they did not replace everybody else. We can only imagine their surprise when they encountered other human communities. What did they make of the differences? Did the anatomical differences jump out at them? How different were their cultures in terms of tools or behavior? When we call the newcomers "modern," we need to avoid thinking of the other human groups as archaic in terms of their level of sophistication in their tools, variability in their cultures, or in the size of their brains. Encounters must have been dramatic events. These diverse human communities had lived in relative isolation, cut off from each other for tens of thousands of years, sometimes even for hundreds of thousands of years. In their relative isolation, they adapted to local or regional challenges.

But their isolation was only relative, and occasional encounters occurred when one group traveled or another pursued game into new areas. At all times, they are best seen as one interlinked population. They were diverse, but not so much than any one group was human while other groups were not. It is true that a comparatively modern variant of humanity appears distinctly in Africa and spreads out. But it is wrong to think that this version of humanity is a new, isolated species that completely replaces all other human species still in existence at the time.

Behaviorally Modern

When it is applied to differences in human anatomy, the term "modern" can be useful as long as we avoid the common confusions and exaggerations. Applied to difference in culture and behavior, "modern" takes on other meanings and refers to such things as cultural complexity and developmental lifespan. These, too, can be helpful concepts if we are careful not to read too much into them.

How do we define "culturally modern humans"? Most often modern behavior is connected with the emergence of symbolic thought. In this view, any behaviors that require symbolic thought are identified as modern. The problem here is that like just about everything else in human origins, symbolic thinking in its simplest forms seems to have arisen early in the appearance of the genus *Homo*.

For example, it has been suggested that the ability to make Acheulean tools requires the ability to visualize a goal and execute a process to achieve the goal. Anyone who removes the flakes from a carefully crafted Acheulean stone ax, for example, must have some ability to see a mental image of the tool before the first blow is struck. Others might agree with the importance of symbolic or representational thinking in Acheulian technology. But is it symbolic enough to be called "modern"? Perhaps a better marker for symbolic culture is the first art. The oldest art first appears much more recently than Acheulian tools. But it is not exactly clear where or when our ancestors became artists or creators of symbolic objects. The most dramatic ancient cave paintings date from 17,000 to about 37,000 years ago, but the very first examples predate 40,000 years. Some recent findings suggest that 100,000 years ago or earlier, human ancestors were making pigments, the key ingredient for painting.

Other experts criticize the standard view of behavioral modernity as a not much more than a thinly-veiled projection of Eurocentric cultural pretensions. If ancestors are like us culturally, they are modern. According to the criticism, the word "modern" means "like us." The more these ancestors had tools and behaviors like ours, the more they were modern. Complaining about this bias, some experts argue that the whole idea of "behavioral modernity" should be dropped in favor of a more neutral standard of "behavioral variability." They accept that it is important to distinguish levels of behavioral or cultural sophistication. But the evidence for what is sophisticated must not be based on how much they are like us. The focus should be how much variation and adaptability there is in the culture of a given community.

For example, how well could some of these ancestral communities adapt and adjust to meet the full range of challenges of their setting? If they modified their behavior in response to new circumstances, they showed behavioral variability. To some extent, *H. erectus* showed behavioral variability. As a general rule, ancestral humans became more flexible behaviorally as time went on. When did their variability become "modern"? Some hold that modern variability was present at the dawn of *Homo sapiens*. At right about this time, stone tools are becoming even more complex, suggesting that modern behavior goes back at least to the "earliest *Homo sapiens* populations who lived in East Africa around 200 kya." As we learn more about the complex story of these first humans, we recognize the limits of our knowledge. "How much farther back in time this capacity for behavioral variability extends remains unknown" (Shea 2011).

Another key component of behavioral modernity is having a modern human life history. In the simplest terms, a modern life history is a lifespan that is slow. It is as if somewhere in our ancient past, a transition was made from a "live fast, die young" pattern to one described as "live slow, die old." Life's main milestones do not change. Birth, childhood, parenting, and death are universal features of mammalian lives. But in humans as compared to even our closest kin, just about everything is slowed down. Compared to our primate relatives, human pregnancies and childhoods are longer, brains develop more slowly, reproduction is later, and molars erupt later.

In most respects, we humans live our lives slowly. But not in all respects. When it comes to weaning our offspring, we live fast. At first that seems odd. But from the standpoint of evolution, what seems like a contradiction may make good sense. Early weaning allows for a second pregnancy to begin long before the first child is fully independent. If that were not the case, humans could not produce enough offspring to perpetuate their communities. "Slower maturation translates into later ages for achieving certain developmental milestones, such as the onset of puberty, adolescence, and so forth, and is intricately linked to the ability of mothers to wean offspring earlier and shorten interbirth intervals, thereby increasing fertility by having multiple, overlapping offspring. This 'stacking' phenomenon is only possible because of the lower energetic requirements for fueling growth in slower maturing organisms compared with the tremendous energetic burden mothers would face having to subsidize the growth of fast-growing, multiple offspring" (Schwartz 2012). In other words, the slower growth of the toddler's brain allows the mother to wean the child, begin another pregnancy, and still feed two or even more growing children with hungry brains. Faster weaning fits with slower development as quite possibly the only way for humans to survive as humans.

All this sounds like human evolution only increases the burdens on human mothers. Whatever truth there may be to this, it is also clear that the burden had to be shared by others in human communities for humanity to evolve. Even when growth is slowed, feeding a demanding toddler creates a burden a pregnant mother cannot meet alone. "The advancement of weaning age throughout human evolution coupled with rapid and early brain growth implies a shift in how the rising energetic demands of offspring are met: initially, energetic costs are subsidized completely by the mother but then by members of the social group through the provisioning of weanlings" (Schwartz 2012).

Taken together, a life pattern like this is often called the "modern human life history package." How might it have evolved? Did it come into existence as a complete pattern in a flash or only slowly, piece by piece? Did it occur early after our divergence from the human-chimp last common ancestor, or is it much more recent? One clue to its origins is that any delay in child-bearing can be a selective disadvantage when early death is common from other causes. If most females die of other causes before they reach their mid-teens, our human ancestors would not have evolved a life history in which a first pregnancies typically occurs late in the second decade of life. We can only guess at what might have allowed early humans to live longer on average. It probably had something to do with better food and better group safety—in other words, eating without being eaten.

Of course, it is impossible to read such things as the age of weaning from fossils. Experts separate basic life history transitions, such as the age of weaning or of sexual maturity, from what they call "life history-related variables," such as the size of the brain or the body or the pace of dental development. These related variables are evident from the fossils, and so they are used to make inferences about the actual life history events. Using this approach, researchers are finding that the modern life history package has evolved piece by piece. Larger bodies, then larger brains, then later dental development seems to be suggested by fossils from *Homo erectus* to *heidelbergensis* and later. This suggests that the modern life history package is not really a package at all, at least not one that arrives all at once. The evidence undermines any notion a dramatic break from archaic to modern life history or that there is "a point in evolutionary history when all these variables switched simultaneously from their primitive non-modern human condition to the modern human condition. The reality seems to be more complicated" (Robson and Wood 2008).

Chris Stringer makes the point clearly: "'Modernity' was not a package that had a single African origin in one time, place, and population, but was a composite whose elements appeared, and sometimes disappeared, at different times and places and then coalesced to assume the form we see in extant humans. . . . During the past 400,000 years, most of that assembly took place in Africa, which is why a recent African origin still represents the predominant (but not exclusive) mode of evolution for *H. sapiens*" (Stringer 2014).

Putting together all the evidence, some experts on human origins suggest that the modern life history is not really present at the beginning of the genus *Homo*. If we go back some two million years or more, we would probably find our human ancestors living a life somewhat more on the pattern of today's chimp than on the pattern that we see as typically human. Over time, a change occurs. But early in our history as a genus, the various components of the modern human life history seem to have come into existence slowly over time, well on their way by the time of *Homo erectus* but not entirely present until much later.

To make matters more complicated, there is even evidence to suggest that our human ancestors may have lived their lives using several different versions of a modern life history. In other words, there seem to have been more than one way to "live slow, die old." While there is a slow transition underway from fast to slow, it seems that a slower lifespan comes in more than one version. In terms of their life histories, some of our ancestors were not like *Australopiths* and not like us, either. It seems instead that they had their own ways of being human and of living human lives.

Humanly Modern

What is the relationship between anatomical and behavioral modernity? There does seem to be a connection between the two. Anatomically modern humans are making their appearance at roughly the same time that a cultural and behavioral transformation is starting to take place. Did anatomical or biological change trigger behavioral change? Or did changes in behavior come first? We will probably never know the full story, if for no other reason than the simple fact that the most interesting behavior changes, such as language and thought, leave no direct evidence in the fossil record. Even so, we cannot help but ask how we became modern humans.

The new or modern humans first appeared in East Africa between 150,000 and 200,000 years ago. After spreading into other parts of Africa, they make their way into Asia and Europe. The most recent evidence suggests that they make their way from Africa to Eurasia in at least two distinct waves (Tassi et al. 2015). The first wave probably took a southerly route, crossing from Egypt to the Arabian Peninsula not too long after they first appear in Africa (Armitage et al. 2011). Some evidence suggests that groups of modern humans were expanding out of Africa as early as 130,000-150,000 years ago. No fossils have been found, but there is a trail of sorts left behind in the form of stone tools. The tools and flakes most resemble tools being made in Egypt about the same time. Whether these tools were made by modern humans making their way across the Red Sea to Arabia is a matter of debate.

Other evidence also supports the idea that the first modern humans to leave Africa took a southern route, maybe following the south coast of Asia from the Persian Gulf through India to Indonesia. Some findings suggest that they were present in these parts of Asia long before they arrived in Europe. In 2015, researchers working in China announced their discovery of modern *H. sapiens* fossils dated between 80,000 and 120,000 years ago (Liu et al. 2015). The claim is controversial, in part because it is so much earlier than most experts expected. If it stands up to scientific scrutiny, it seems to confirm the idea that the first modern humans left Africa at least 100,000 years ago and followed a southern route, keeping to warm climates and avoiding Neandertals who lived further north across Eurasia.

This evidence from China may fit together with other evidence of *Homo sapiens* from southern Asia. The earliest modern human presence in Australia and Melanesia seem to go back to a time well before similar humans reached Europe. Whether the earliest *Homo sapiens* in Australia and Melanesia were part of the same wave that settled southeast China seems plausible. They must have gotten an early start if they reached New Guinea by 50,000 years ago and Australia by 45,000. If so, then it seems likely that these first modern humans to reach this part of the world left Africa in an early wave of migration, following a southern route that led eventually down through modern Malaysia to Indonesia and beyond (Reyes-Centeno et al. 2015). During the low sea levels of the Ice Age, dry land was open most of the way.

Evidence also suggests that at just about the same time, early *Homo sapiens* traveled down the Nile and crossed the Sinai. They entered the Levant and seem to have lived in the northern part of modern Israel. Remains dating from 90,000 to 130,000 years ago have been found. No one knows exactly what happened to these early modern humans. They seem to have disappeared. Even more intriguing is that they lived in caves previously occupied by Neandertal, who seem to have returned after the modern humans were gone. Some experts describe this expansion as a "failed colony." Evidence suggests that a later wave of modern humans, or maybe a series of them, returned to the Levant roughly 56,000 years ago, following a natural pathway along the eastern end of the Mediterranean Sea. Eventually they reached southern Europe a little more than 45,000 years ago.

When we put all this evidence together, we can see that there is a growing case to be made for the idea that modern humans left Africa in successive waves. These waves or groups may have been separated by as much as 50,000 years. If so, then we have to ask whether the people who make up the successive waves were slightly but noticeably different from each other. Based on the evidence available today, researchers can only begin to ask the question. Were the waves different from each other, and do the differences show up in subtle ways in the DNA of human populations today? A recent study argues that there indeed is a small but detectable difference based on early migrations. Those whose ancestors lived in Australia and Melanesia have slightly different genetic profiles from mainland Asians, for example. A research group led by Francesca Tassi describe the difference this way: "We conclude that the hypothesis of a single major human dispersal from Africa appears hardly compatible with the observed historical and geographical patterns of genome diversity and that Australo-Melanesian populations seem still to retain a genomic signature of a more ancient divergence from Africa" (Tassi et al. 2015).

From Tools to Art

If groups of early *Homo sapiens* left Africa at different times, did they bring different levels of technology and culture with them? During the critical period of 50,000 to 100,000 years ago, cultural and technological advances were taking place in Africa. No evidence of cave painting goes back this early, but other signs of art have been found. The earliest known site for the production of the pigment ochre, for instance, dates to 100,000 years ago. In a cave at the very southern tip of Africa, researchers have found remains of a complex ochre-processing facility. Later, decorative beads appear in Africa in the south as well as in the north.

Significant advances in stone tools go back even further. New evidence suggests that in South Africa, well before "modern" humans arrived, earlier humans had the ability to attach stone tips to spears. The process is complicated. Specialize tips must be created, and then some sort of adhesive is needed to attach them to spears. Recent discoveries suggest that humans were able to do this 500,000 years ago (Wilkins et al. 2012).

Moving forward to a time about 100,000 years ago, it seems that the world's technological hotspot is once again in South Africa (Douze et al. 2015). One research team comments on new findings this way: "Recent archaeological data also showed that modern behavior (such as symbolic culture and complex tool production) arose at a relatively early stage of human evolution, contrary to prior studies that argued for the later development of complex cognition ~45 kya" (Campbell et al. 2014). We may never know for sure, but some evidence suggests that language arose first in Africa and traveled with those who left. "Thus, key behavioral and morphological traits that define modern

Homo sapiens may have evolved fairly closely together in Africa over the last 200,000 years" (Campbell et al. 2014).

Until recently, many experts thought that human culture went through a sudden transition to a modern, artistic, and symbolic level. Some researchers even spoke of a "cultural big bang" or a "cultural revolution," referring to advances in Europe beginning 35,000 years ago. The most dramatic evidence to support this view is the cave art of Western Europe, especially in modern France and Spain. These paintings show that cultural advances are occurring in Europe at about this time. But what happens in Europe is part of a larger and slower pattern, centered mostly further south and going back to a time before Europe is occupied by *Homo sapiens*.

The essential point here is not simply that Africa is the location of key cultural advances made by *Homo sapiens*. The critical issue is that these advances reach further back in time than we once thought. Modernity did not happen in a flash. Our ancestors had to work at it, or so it seems. Like human anatomy itself, behavioral and cultural modernity is a complex package. It came into existence piecemeal, not as a whole and certainly not in a singular "big bang." Anatomically modern humans do not arise in a single flash of evolutionary advance. In the same way, behaviorally modern humans arise only slowly, as small innovations build on themselves. These two dimensions of being modern, modern anatomy and modern behavior, begin to arise together in Africa. Most of the key steps occur in Africa before modern humans migrate into Eurasia.

Recent discoveries are now showing that the creation of art reaches back earlier than we once thought. About 70,000 years ago, human communities along the North African Mediterranean coast were making beads of shells. Small seashells, carefully drilled so they can be strung together, have been found many miles away from the shoreline, suggesting that they were valued and possibly traded (d'Errico et al. 2009). From elsewhere along North Africa, similar findings have been report (Vanhaeren et al. 2006).

At about 70,000 years ago, however, the southern end of Africa seems to be the global epicenter for technological and cultural breakthroughs. There are no dramatic cave paintings going back this far, but tool making was highly sophisticated. Especially notable were the fine, small tips designed for mounting on projectiles. Could these be the first arrowheads, designed for use with a bow? Perhaps they were lightweight spears, hurled at great distances with an atlatl or spear-thrower. In the conclusion of their report, researchers point out that "early modern humans in South Africa had the cognition to design and transmit at high fidelity these complex recipe technologies." The somewhat sobering consequence was "increased killing distance and power over hand-cast spears." Hunting trips were probably more successful than before. But with their new weapons, these more modern humans were also extending "the effective range of lethal interpersonal violence" (Brown et al. 2012).

It is hard to imagine that the creators of this advanced technology could make it or use it without language. Tools and weapons like these have to be made just the right way, following a series of complex steps with a clear goal in mind. This seems to require "high fidelity transmission and thus language" (Brown et al. 2012). Once it was thought that culture so complex first appeared in Europe beginning around 40,000 years ago. But now we know that these tools first appeared in South Africa. In fact, the latest evidence suggests that some of these technological and cultural advances first appeared far earlier, with key steps first appearing more than 100,000 years ago. Then, beginning around 71,000 and lasting for some 11,000 years, a culture capable of producing advanced tools was established on the Cape. Experts tend to refer to this as the Pinnacle Point culture.

A little further east along the Cape and going further back in time, researchers have found an early engraved object. The piece that was found is only a fragment. But the few inches of stone seem first to have been polished, then carefully scratched with a stone tool. The straight lines are mostly parallel to each other, while some are diagonal. Traces of ochre-like substance have been found in the cut lines. The experts who describe it say that it "may represent one of the oldest instances of a deliberate engraving" (d'Errico et al. 2012), unless of course we take into account a possible engraved object created by *Homo erectus* on Java, reported in 2015 (Joordens et al. 2015). The South African object, found in a cave along the Klasies River, is dated to 85,000 to 100,000 years ago. In all likelihood, other objects like this are waiting to be found. Only then can we get even a glimpse into the lives and culture of the people who made this Klasies River engraving, or how they may be connected to the Pinnacle Point culture. Some evidence suggests that the manufacture of the pigment ochre was known throughout this region before 160,000 years ago. If they were not yet painting on caves, these early humans seem to have painting something, perhaps their own bodies.

Cave Paintings and Musical Instruments

About 40,000 years ago, these ancient people dipped a hand in paint and pressed it against cave walls. Or they created stenciled handprints by blowing paints from their hands or mouths in a kind of aerosol spray, leaving behind a painted haze around the clear, blank space where their hand was held. Hand prints dating from stone age cultures have now been found on cave walls around the world. Recent advances in dating techniques are making it possible to ask new questions about the first artists. Precise measurements of some of the oldest hand stencils in France and Spain have also revealed that the persons who made them were "predominantly female" (Snow 2013).

The most intriguing question of all, of course, is where painting began. The location of the earliest art is the key to knowing something about the first artists. Who were the first humans to express themselves in such an enduring, symbolic, and creative way? Of course, the ancient cave paintings in France and Spain are the best known. But the most recent evidence points to Indonesia as the possible location of the world's oldest art. On the Indonesian island of Sulawesi, hand stencils have been dated to at least 39,900 years old, just older than their European counterparts. An amazingly realistic painting of an animal, the now extinct "deer-pig," dates back 35,400 years (Aubert et al. 2014).

As researchers apply the new dating techniques to known cave art, claims for the oldest art may need to be changed. But the fact that art that appears on the remote island of Sulawesi predates or is roughly simultaneous with similar art in Western Europe raises perplexing questions. How closely related were these two human communities? They probably left Africa in different waves. Those making their way along the southern edge of Asia and all the way to remote islands probably left first, maybe even as early as 80,000 to 100,000 years ago. Those in Europe probably left Africa in a later wave, roughly 55,000 years ago. Even so, at virtually the same time, they seem to have started painting on cave walls. How can we possibly explain why this happened? It seems highly improbable that art was invented in one place and taken to the other in so short a time.

Maybe the explanation is that these human communities brought art with them from Africa. In that case, Indonesia is not really the site of the world's first art, and neither is Western Europe. The humans migrating to these places brought art with them. The key problem with this idea is that there is no evidence of these humans having painted anything in Africa or along the way to their new homes. Another problem is that if they brought art with them, art has to go back roughly 100,000 years. There is no evidence to support this, but then there is no other explanation for how art shows up in two highly remote locations at roughly the same time. For now, the origin of human symbolic culture is hidden from view.

There is more clarity, at least for now, when it comes to the question of the world's first musical instruments. The oldest known instrument is a flute made from the bone of a bird. It is believed to be 42,000-43,000 years old, dating back to a time when modern humans were just entering Europe (Higham et al. 2012). The site of discovery is in southern Germany, suggesting that humans were using the Danube River as their path of access into northern Europe. The flute was found in a system of caves where figurines have also been found. Other flutes almost as old have been found throughout the same region. Bones from birds, already hollow on the inside, were carefully drilled so that different pitches could be played, much the way we do today with flutes and musical whistles.

From roughly the same time come beautifully carved ivories shaped like birds or other animals. And most dramatic of all are the paintings found deep in caves in modern France and Spain, in places like Chauvet Pont-d'Arc, where the first complex artistic compositions appear starting about 37,000 years ago. When we see them today, we are struck not just by their skill and beauty but most of all by the recognizable humanity of the artists.

Some have speculated that religious yearnings or states of awareness are present at the birth of complex art. We can only wonder about the thoughts, fears, or dreams of these first artists. Their mental states are forever hidden from us, but in the art they left behind we can see clear evidence of something new emerging. Here is a human community creating art, using symbols perhaps to communicate, perhaps to define or assert themselves, perhaps to form a connection with a realm beyond ordinary objects. We will return to some of these themes in Chapter 9.

By today's standards, the pace of cultural change is still very slow. Agriculture, which arose independently in many communities around the world, only appears between 10,000-15,000 years ago. Before agriculture, humans are hunter-gatherers, getting their food from a wide range of animal and plant sources, including the wild versions of grains that are later domesticated. Settlements were small, scattered, and temporary, as human groups followed their food from place to place and season to season. All these factors make sophisticated cultural development difficult. With the advent of agriculture, culture-making itself seems to change. Permanent settlements with larger populations give rise, in time, to specialized crafts. Successful agriculture means more people, and eventually there is an expansion of farmers from Middle East into Europe. Population growth triggered waves of migration from today's Middle East into Europe, where the newcomers largely replaced the scattered communities of hunter-gatherers (Omrak et al. 2016). In the Middle East, population growth gave rise to cities, institutional religion, class and caste structure, armies and empires, and written records

.

7

Born Again Neandertals

Early *Homo sapiens* spread from East Africa some 100,000-150,000 years ago. As they made their way across the African continent and then into Asia and Europe, a big surprise was waiting for them. They were probably looking for elephants and other large game. What they found instead were other humans not very different from themselves.

Unfortunately, we do not know very much yet about the various forms of humanity that lived in Africa around 200,000 years ago. By contrast, we know a good bit more about the ones living in Europe and Western Asia. We know they lived in small settlements from Spain all the way east to the Ural Mountains. When the climate changed and the ice in the north forced them south, they sought refuge along the shores of the Mediterranean. In warmer periods, they occupied lands as far north as the edge of the Russian Arctic. These human communities were diverse in culture and in anatomical traits. And like all living things, they changed over time. To us today, however, they are known simply as the Neandertals.

When the early *H. sapiens* newcomers entered Asia and then Europe, the Neandertals had been living there for hundreds of thousands of years. In many ways, the African immigrants and their northerly cousins would have looked very much alike. It is likely that the newcomers had darker skin, stood a bit taller, and were less robust. More than anything else, their faces might have seemed strange to each other. Where the Neandertal face is forward and has a heavy brow ridge above the eyes, the immigrants' faces seemed flatter than the faces of Neandertals, who lacked chins. But after hundreds of thousands of years of surviving in the cold, Neandertals were adapted for life in Eurasia.

When they made contact, what did they think of each other? Did they greet each other with fear, curiosity, kindness, or violence? Probably some of each. Mostly, they probably kept away from each other.

But we know for sure that they met. The evidence of their encounters is carried in the DNA in every cell of our bodies. Most human beings alive today carry the legacy of those ancient sexual encounters that produced at least some viable offspring. Despite their biological differences, Neandertals and the early *Homo sapiens* immigrants interbred successfully, and we are their distant offspring. Of all the recent scientific discoveries that have shaken up what we once thought we knew about human origins, nothing has been quite as provocative as the discovery of Neandertal genes in our DNA.

What practical difference does it make? Does having Neandertal DNA change the way our bodies or brains function? Scientists are hard at work trying to answer that question. But the significance of having Neandertal DNA goes far beyond any practical biological consequences. It changes the way we see Neandertals, and it changes the way we see ourselves. Suddenly, Neandertals are no longer mere curiosities or the brunt of "caveman" jokes. Without them, we would not be what we are today.

Who Were the Neandertals?

The first Neandertal bones were found in Belgium in 1829. At that time, no one knew what to think of them. The unusual bones were probably best explained by disease, the experts said. In 1856, more bones were found. Not knowing what else to call them, people started calling them "Neandertals" after the valley in western Germany where they were found. The valley itself was named in honor of a much-loved 17th century Protestant pastor and hymn-writer, Joachim Neander, who enjoyed walking there. In an ironic bit of historical trivia, this means that the first discovery of another form of humanity was named for a theologian. Since the mid-1800s, hundreds of other samples of Neandertal remains have been found across Europe and western Asia.

Popular reaction to the Neandertals is a story in itself. At first, they were seen as extremely "primitive," largely ape-like or at least very "un-European." Just at the time when Europeans were solidifying their colonial empires and justifying their domination of other peoples, European Neandertals came to symbolize primitive humanity. All too often these notions came together in a simplistic, self-congratulatory comparison between modern European or civilized humanity and those primitive, not-fully-evolved forms that lived long ago or far away (Wolpoff 2009; Wolpoff and Caspari 2011).

Today, experts disagree on the best way to classify all the Neandertal findings. Neandertals changed over time and differed over the full extent of their range, which is now known to extend well into Asia. Depending on how they are defined, some version of "early" Neandertals first appeared about 500,000 years ago. The most characteristic form of Neandertal appeared more like 200,000 years ago and survived at least until 40,000 years ago and maybe even later. They ranged from Gibraltar east to Siberia and at least to 50 degrees latitude north. Their population size was probably never very great, numbering in the tens of thousands or slightly more.

As much as they were like us, they were also different in important ways. Compared to us, they had heavy bones and strong muscles. Where our ribcage tends to broaden at the top, enhancing upper body strength, the Neandertal ribcage was broader at the bottom, maximizing core strength and a lower center of gravity. In particular they had a heavy, continuous, double-arched brow ridge across the face above the eyes. On the whole, their mid-face was projected forward more than ours, accentuated by a receding jaw and the lack of a projecting chin. Their teeth were larger than ours. Overall, compared to ours, their skulls look elongated from the side.

And compared to ours, their brains were slightly larger, about 1520 cc on average. New evidence also suggests that in the first year of life, the cranium of the Neandertal child developed differently from ours. On the whole it seems that their children grew more quickly. According to one study, "the developmental patterns differ markedly in the period directly after birth: within the first year of life, only modern human endocasts change rapidly from an elongated to a more globular shape" (Gunz et al. 2010).

Were the Neandertals a species separate from *H. sapiens*? Many accept this idea. It is certainly easier to speak that way, referring first to Neandertals and then to modern human immigrants as two distinct groups. But does the difference really justify calling them a separate species? Anyone who thinks that individuals from separate species cannot produce viable offspring will have to say that we must belong to the same species. After all, today we know that Neandertals and moderns interbred successfully. That alone is proof for some that even though they may be distantly related, Neandertals and modern humans are members of the same species. But is it right to think that members of different species cannot reproduce successfully? Most experts reject that definition of species. For them, it makes the most sense to describe Neandertals as a distinct species, but one that is closely enough related to modern humans so that viable interbreeding can occur.

The question of species is tied closely to the question of where Neandertals came from. When did the Neandertal line diverge from the line that led to modern humans in Africa? No one knows for sure, but evidence suggests that the Neandertal/modern divergence goes back more than 500,000 years. Whether the last common Neandertal/modern ancestor is best labeled *Homo antecessor* or *Homo heidelbergensis* is a matter of debate. The key point is that Neandertals and moderns share a common ancestor, then diverged to become two distinct populations or species, then converged.

During times of lineage separation, evolution has the opportunity to run on at least two tracks at once. Then, if the lineages converge, anything of lasting benefit that has evolved separately might be brought together. This pattern of divergence and convergence or separation and hybridization is critically important to the full story of our origins. It happened with modern humans and Neandertals, not just once but many times. And it happened with modern humans and other earlier forms of humanity. Separation allows evolution to proceed locally. Hybridization opens the possibility that anything beneficial might be added to a new, converged community.

Neandertal Culture

Biologically, the modern humans from Africa and the Neandertals in Western Asia and Europe were overwhelmingly similar and yet distinguishable when they met some 50,000 years ago. In terms of cultural sophistication, how did the two groups compare? The new arrivals from Africa brought techniques and cultural patterns that were different from Neandertal culture. But were the anatomically modern humans more advanced than the Neandertals? Once we all thought so. After all, what were these Neandertals if not "cave men," hopelessly unintelligent and primitive? In fact, we thought, they died out biologically because they were behind the times culturally.

Now we know otherwise. The most recent research shows that the two groups were largely similar in terms of sophistication. The two types of humans were "very similar in terms of what were once thought to be standard markers of modern cognitive and behavioral capacities, such as diversity of subsistence strategies and diet, use of minerals, use and transport of lithics, shells, personal ornaments, and hafting [attaching stone tips to spears], and pyrotechnology" (Roebroeks and Soressi 2016).

What was life like for Neandertals? What did they eat? Did they create art or decorate themselves with pigments or jewelry? Did they bury their dead? Not so long ago, Neandertals were seen as the very definition of human primitiveness. We thought they had no capacity for culture or innovation. We pictured them as brutes who communicated by gestures and grunts and whose idea of advanced technology was to hit things with big rocks. In recent decades, that view has changed. We now see them as generally on a roughly equal footing culturally with other human groups in other parts of the world existing at the same time.

Before the arrival of modern humans, Neandertals were making sophisticated tools, and at least some of them were decorating themselves. The term "Mousterian" is often applied to Neandertal tools and culture. Mousterian tools are found throughout Europe from 600,000 to about 40,000 years ago. Recent evidence also show that Neandertals made advanced use of fire, not just for warmth or to cook their food but as component of their technology. They were able to create an adhesive of birch bark pitch by carefully heating several ingredients. This was used to fasten stone tips to spears. In all these ways, Neandertals were quite sophisticated in use of tools, close to or pretty much at par with what is happening elsewhere at, say, 50,000 years ago or earlier.

There is no reason to doubt that Neandertals used language at much the same level as any other human population at the time. One specific gene that seems to play a role in language, the FOXP2 gene, is now known to have been present in Neandertals in a form like our own, with some subtle variations (Maricic et al. 2012). That alone does not prove that Neandertals did speak. The current evidence suggests that they probably were capable of some sort of speech, but experts debate whether they were capable of the complex syntax and large vocabulary that is common among humans today. Speech is a complicated process that involves brain, mouth, and throat. In all these dimensions, Neandertals were enough like us that it is hard to imagine that they did not communicate verbally at all. But there is no direct evidence to prove that they talked, and certainly nothing to suggest that they spoke as we do.

For reasons not yet fully understood, the modern human immigrants coming into Eurasia seem to make advances in culture and technology while the Neandertal show signs of lagging behind. Both groups were sophisticated and able to create new techniques. But whether the newcomers brought new ideas with them or learned them from later waves of immigrants, the modern humans entering Neandertal regions seem to enjoy a technological and cultural advantage. This is especially true as they entered Europe, starting at around 45,000 years ago.

During this time, Neandertal culture is not entirely static. Among experts, there is a sharp debate over the best way to understand changes in Neandertal culture that occur in Europe after their encounter with the modern immigrant some 45,000 years ago. Were they learning new techniques from the modern humans? Or were they suddenly making their own discoveries and advances? The debate centers on several specific sites and how best to understand the local population fluctuations that were going on throughout Europe at this critical transitional time. In some cases, the same sites were occupied first by Neandertals and then by the newcomers. Who created the tools that were left behind?

The technological advantage brought in from Africa was not extreme and should not be overstated. One independent Neandertal advance, for example, involved making tools not just from stone but also from animal bones. It was shown in 2013 that the oldest specialized bone tools made in Europe were made by Neandertals. Similar tools were being made in Africa even earlier, but the African immigrants had not yet reached Europe when these tools were made about 50,000 years ago. The bone tools were used to treat animal hides by softening and smoothing them, making them tougher and more waterproof. The tool, called a *lissoir*, was made by grinding and polishing sections of animal rib bones. The newly discovered *lissoir*, found in southwestern France, is dated to just over 50,000 years ago. At that date, experts cannot entirely rule out the possibility that Neandertals learned how to make these tools from the first modern humans to reach Europe. But it seems more likely that the Neandertals invented these techniques on their own (Soressi et al. 2013).

In 2015, researchers took a closer look at eagle talons that were discovered about 100 years ago in what is now Croatia. Based on their analysis, the team concluded that the talons were collected by Neandertals about 130,000 years ago, most likely for decorative purposes in the form of a necklace or a bracelet. They are the earliest evidence for jewelry in the European fossil record and demonstrate that Neandertals possessed a symbolic culture long before more modern human forms arrived in Europe" (Radovčić et al. 2015). Experts debate just how much the Neandertals were capable of symbolic culture. If anything, the evidence is shifting toward the idea that when it came to complex thinking, they were about as advanced as any other humans living at roughly 50,000 years ago (Caron et al. 2011).

Just as the modern humans were entering Europe, cave art appears. The earliest paintings in Europe date to 40,300. There is evidence that well before that date, Neandertals were making ochre, a widely used prehistoric pigment. In fact, the use of ochre by Neandertals in Europe and by pre-modern forms of humanity in Africa goes back perhaps as far as 250,000 to 300,000 years ago. No paintings go back that far, so no one knows for sure how ochre was being used. Europe's oldest paintings, located in a cave in northern Spain, consist of circles or large dots on a cave wall. At 40,300 years ago, it seems most likely that these circles were painted by the new immigrants. But this is almost exactly the time that Neandertals were disappearing from that region and modern humans were arriving (Higham et al. 2014). As a result, experts cannot say for sure who painted the circles.

Close to the end of the period of Neandertal presence in Europe, a group was living along the Mediterranean coast at Gibraltar. They lived in Gorham's Cave, where they made engravings on a large stone. This is the first known human engraving made in a dwelling (Rodríguez-Vidal 2014). Compared to later cave paintings, these early marks are simple, almost nothing more than scratches. But they seem to suggest that the Neandertals who made them were trying to express themselves in some way. The engravings are older than 39,000 years ago, made while Neandertal but not modern humans were living in the area.

According to the team of experts who described them, "The engraving at Gorham's Cave represents the first directly demonstrable case in which a technically elaborated, consistently and carefully made nonutilitarian engraved abstract pattern whose production required prolonged and focused actions, is observed on the bedrock of a cave." Using stone tools, they cut straight lines into the bedrock to make a pattern that resembles a complex version of the game "tic-tack-toe." Those who made the discovery "conclude that this engraving represents a deliberate design conceived to be seen by its Neanderthal maker and, considering its size and location, by others in the cave as well. It follows that the ability for abstract thought was not exclusive" to modern humans (Rodríguez-Vida et al. 2014). It does not appear that these marks at Gibraltar were painted. It does appear, however, that this is the world's oldest engraving that was incorporated into a space for living.

Did Neandertals bury their dead? The answer depends in part on what we mean by "bury." It is one thing to put a body into a hole and cover it with stones or dirt. It is quite another thing for a community to do this repeatedly over time and to do it with a certain intentionality. The evidence is pretty clear that Neandertals covered bodies. It is less clear that they did so as a part of their culture. At the center of the debate is the site of La Chapelle-aux-Saints in western France. Recent analysis of the site suggests that a complete body was buried rapidly and with enough care to support the idea that the burial was intentional (Rendu et al. 2013). If so, then in terms of how they treated the dead, the cultural differences between Neandertals and the modern human immigrants were indeed not very significant. Based on the interpretation of sites like La Chapelle, our assessment of the cultural complexity of Neandertals and their capacity for symbolic thought is clearly higher than it used to be.

Perhaps the most astounding Neandertal cultural achievement of all was just reported in 2016. A community of Neandertals living in southern France was venturing deep into Bruniquel Cave. About 1000 feet from the entrance, they built structures using stalagmites that they broke from the cave's floor and ceiling. The structures date to about 176,000 years ago. Working by torchlight, they used about 400 stalagmite pieces like building logs, stacking them to create semicircles and other shapes. They built fires next to the structures. Their work was nicely preserved when the cave itself encased everything in a thin layer of stone. Although there is trace evidence of other, older human structures, these Neandertal "edifices [are] among the oldest known well-dated constructions made by humans." Their existence shows "that humans from this period had already mastered the underground environment, which can be considered a major step in human modernity" (Jaubert et al. 2016; Soressi 2016). No one claims to have any idea why these Neandertals went to such effort to create these structures.

Disappearing or Assimilating?

During the key transitional period between 40,000 and 50,000 years ago, Neandertals and modern humans were both present in Europe and Western Asia. During this time, the Mousterian culture of the Neandertals gradually disappears. In its place is another tool-making culture, the "Aurignacian," which is more variable and sophisticated. Experts agree that Aurignacian culture is exclusively associated with the modern humans from Africa. It seems to show up first in the Levant and then spread with modern humans into Europe and Western Asia.

The challenge here is that not everything fits neatly into the Mousterian or the Aurignacian category. For example, researchers speak of a precursor to the Aurignacian culture, something they call "Protoaurignacian." Who made it? The debate is critical to the broader questions. When did modern humans arrive in various places in Europe? A group of experts led by Stefano Benazzi describes the debate this way: "The Protoaurignacian culture is pivotal to the debate about the timing of the arrival of modern humans in Western Europe and the demise of Neandertals. However, which group is responsible for this culture remains uncertain" (Benazzi et al. 2015). The problem facing experts is that Neandertals were still present in Europe when Protoaurignacian culture makes its appearance. This happened around 42,000 years ago.

For reasons still mostly unknown, Neandertals declined in numbers and ultimately disappeared as a distinct population around 40,000 years ago, with the final remnant groups probably all gone by 30,000 years ago. Did the culture of the newcomers trigger their demise? Perhaps the better tools and hunting techniques of the immigrants meant the Neandertals had less game to hunt or that they lost any inter-group conflicts. Their disappearance was gradual, and evidence suggests that it occurred at different times in different regions. This suggests that there were several causes for their decline.

Or perhaps the Neandertals did not decline so much as they become absorbed into a new community. Some experts argue that rather than just disappearing, remnant Neandertal populations were absorbed or fused with communities of new immigrants. The time of overlap between the Neandertal and the modern populations was 10,000 years or more if we consider Europe and Western Asia as a whole. In any specific region, the overlap was more in the range of 2500 to 5000 years, plenty of time for cultural and biological assimilation to occur.

For example, we know that one route the new arrivals took was up the valley of the Danube. Aurignacian tools that date to 43,500 years have been found as far up the valley as Austria (Nigst et al. 2014). Neandertals were in the same region at the same time. A recent analysis of findings in modern Romania, further downstream, provides a surprising result. If successful interbreeding happened between Neandertals and modern humans, then first generation hybrid individuals with roughly 50/50 DNA must have existed, for instance living in a community that was predominantly modern. But in that case, with each generation after the Neandertal/modern admixture, the percentage of Neanderthal DNA would be cut roughly in half. Finding the bones of an individual close to the moment of interbreeding would be a rarity. And yet this is exactly what researchers claim to have found in a cave in southwestern Romania. The team that extracted the DNA and analyzed it were guided by the experts who already thought the bones were a little unusual.

Even so, according to Svante Pääbo, the leader of the genetics team, "It is such a lucky and unexpected thing to get DNA from a person who was so closely related to a Neandertal." In a comment on Pääbo's institutional website at the Max Planck Institute for Evolutionary Anthropology, Pääbo went on to say: "I could hardly believe it when we first saw the results." Pääbo's team knew which fossils to test because other experts already noticed that this individual was somewhat Neandertal and somewhat modern, showing what experts call a mosaic of traits or features. What the DNA revealed was that this individual's mixed traits were the result of interbreeding that had occurred as recently as 4-6 generations earlier (Fu et al. 2015). The interbreeding probably occurred locally, and it is dated to about 40,000 years ago. It should also be made clear here that a first or second generation hybrid does not necessarily look like something half way between the two parents. From other species, we know that sometimes hybrids look like one parent or neither. Even so, for many years, some researchers studying European human remains dating from 25,000 to 40,000 years ago have claimed that some of the bones show mixed or mosaic traits.

Getting Bones to Talk

The relatively new field of paleogenetics was key to the discovery of the individual in Romania with recent mixed ancestry. But the impact of this new field is far more broad and profound that the discovery of one individual with unusual great-grandparents. Although it is still a new field, paleogenomics has already transformed the science of human origins research. By figuring out how to recover, decontaminate, and reconstruct DNA from fossilized bones tens of thousands of years old, researchers have open a whole new window onto our past. Thanks to this work, we know about the DNA from several Neandertal individuals with the same sort of detail that we know our own.

One big surprise was the discovery of a whole new human species. This simply could not have happened without the ability to recover ancient DNA from fossilized bones. Based on DNA reconstruction alone, researchers have identified a form of humanity distinct from Neandertals and from modern humans alike. The new group of humans are called Denisovans, after the cave in Russia where their tiny fossil fragments were found (Reich et al. 2010).

Once DNA is recovered and sequenced from Neandertals or Denisovans, it can be compared with DNA from living humans or from modern humans that may have lived thousands of years ago. Through detailed comparisons, scientists are able to piece together the most likely lineages and evolutionary relationships. Because paleogenomics is still a comparatively new tool for human origins research, more findings are sure to be made and more surprises are certainly on their way. Technical advances will also be made, probably pushing back the age of the bones from which readable DNA can be extracted. As of 2016, the oldest human DNA that has been reconstructed dates to 430,000 years (Meyer et al. 2016).

Already, enough has been learned using this new technique to force experts to revise the story of recent human origins. The new account is a modified version of the "Out of Africa" theory. Today, experts tend to agree that anatomically modern humans began to emerge probably in East Africa some 200,000 years ago, probably in part because of the way various lineages diverged and converged. From that starting point, small groups of these comparatively more modern humans began to move across Africa. Evidence suggests that as they traveled south and west, they encountered and interbred with other groups (Hammer et al. 2011).

Sometime about 80,000 years ago, a small band of modern humans made their way out of Africa, probably taking a southern route across the south end of the Red Sea to the Arabian Peninsula. They continued moving east across southern Asia. Somewhere along the way, probably on islands in present day Indonesia, some of them interbred with the mysterious Denisovans. Compared to other human groups alive today, Melanesians carry the most Denisovan DNA (Reich et al. 2011; Meyer et al. 2012). It is clear that the Denisovans also interbred with other archaic humans. A really surprising finding was reported in 2014, when researchers in Spain showed that pre-modern humans living there about 300,000 years ago showed clear signs of Denisovan ancestry, in spite of the fact that they lived so far away from the Denisova Cave and even further from the Melanesians (Meyer et al. 2014). New findings may clear up today's questions, but they are also likely to raise new ones.

Traveling in small groups numbering a few hundred to a few thousand, later waves or bands of modern humans made their way out of Africa. One group lived for a while in the Levant, then disappeared. Sometime between 50,000 and 60,000 years ago, a more successful wave moved into that area, where Neandertals also lived. One place where interbreeding probably occurred was in the area around Mount Carmel in present day Israel. Experts estimate that the date of this interbreeding was most likely between 47,000 and 65,000 years ago (Sankararaman et al. 2012). In a community made up of the modern humans, some intermixed or hybrid individuals also lived, and at least some of them were able to reproduce. Over time, a small but significant portion of the gene pool of that community came directly from Neandertals. This seems to be the best explanation for why Neandertal DNA is a small but evenly distributed part of the genome of people with European and Asian ancestry.

Today it is clear that interbreeding between modern humans, Neandertals, and Denisovans occurred over and over, probably from 100,000 up until 40,000 ago and possibly even later. One recent study shows how the DNA of modern humans contributed to the DNA of the Neandertals (Kuhlwilm et al. 2016). In other words, the exchange of DNA went both ways. It has also been shown that the modern humans who lived in Europe about 45,000 years ago carried Neandertal DNA in their genomes at a level of 3-6%. Today, the level is more like 2%. We also know that this original "modern" European population has effectively disappeared. For the most part, Europeans today are descended from newcomers who arrived about 14,000 years ago from the Middle East. This suggests that human populations are dynamic and that "turnover and migration have been recurring themes of European prehistory" (Fu et al. 2016).

As scientists sample more genomes from living individuals and compare them with reconstructed DNA from the past, they are getting a more complete picture of these interactions (Vernot et al. 2016). Even so, it appears that on the whole, Neandertal DNA is not particularly useful to later humans. There are, however, a couple of exceptions to this general rule.

The Legacy of Interbreeding

Does ancient interbreeding have any practical consequences? Recently, evidence seems to be building that Neandertal DNA helped modern humans survive in Eurasia (Simonti et al. 2016). More than anything else, the Neandertal DNA present in today's people of European and Asian origin tends to be related to innate or inherited immune responses (Deschamps et al. 2016). Neandertals went through the slow process of acquiring these genes through their evolution during a time of separation. When they met modern humans, these particularly useful genes were transferred. Procreation between the two groups, whenever successful, resulted in what is often called "gene flow." Useful genes pass from one lineage to the other, along with genes that were not so useful or may have been harmful. Over time, natural selection tends to keep what is useful.

Svante Pääbo, one of the leading developers of the technology of ancient DNA retrieval, describes adaptive introgression this way: "If favourable alleles have emerged in one group, they can spread to other groups relatively rapidly by gene flow. This process [is] called 'adaptive introgression'." What useful genes might have come from Neandertals to modern humans? Pääbo answers this way: "Not surprisingly, genes involved in systems that interact directly with the environment such as the skin, immune system, digestive system and intermediary metabolism are those that are most commonly found to show evidence of positive selection and also those in which adaptive introduction of alleles from now-extinct hominins seems to be the most frequent" (Pääbo 2015).

Compared to modern humans that lived in Europe close to 40,000 years ago, today's Europeans have somewhat less Neandertal DNA. In particular, the Neandertal version of the Y-chromosome, found in males but not in females, seems to have disappeared. In its Neandertal form, the Y-chromosome has never been found (as of 2016) in a modern human. One reason is that it seems to contain genes that trigger an *in utero* immune reaction from a human mother who might be carrying a male child with the Neandertal version of the Y-chromosome (Mendez et al. 2016).

For now, our knowledge of interbreeding is largely limited to modern human interaction with Neandertals and Denisovans. Other interbreeding probably occurred. Most experts are now beginning to suggest that whenever diverging communities reunited, interbreeding was probably the most common outcome. This was very likely the case wherever modern humans traveled, whether across Africa or into Eurasia (Hammer et al. 2011; Triska et al. 2015). Divergence and hybridization "is the rule, not the exception, in human evolution" (Ackermann et al. 2015).

Why is interbreeding so important to our understanding human evolution? In part it is because gene flow plays a critically important role in bringing us into existence biologically. At the very least, gene flow speeds up evolution. But more than that, hybridization multiplies the possibilities for what might evolve. One team of experts summarizes recent findings with this suggestion: "These emerging studies raise the possibility that hybridization played not merely a small or ephemeral role, but a central role in the emergence and evolution of *Homo sapiens* through the introduction of new variation, and production of new genetic amalgamations and innovation, thereby opening up more evolutionary possibilities" (Ackermann et al. 2015). In other words, we should not think that well-defined modern humans just happened to interbreed with a few Neandertals. It seems more accurate to say that interbreeding all along the way is what helps define modern humans.

Human Emergence through Interbreeding Lineages

If this is true, then it changes fundamentally the way we understand our origins. Even today, some who accept the idea that humans evolved tend to think we modern human beings came into existence suddenly as a unique species. Such a view bears an odd resemblance to the older idea that human beings were created directly in our current form. It now appears that we were not created, nor did we evolve, directly into our current modern form. Today's humans came into existence through a process that is not only long and complex but characterized by divergence and convergence. Our ancestral communities tended to go off in different directions, geographically and in terms of anatomy and behavior. Then we reconnected, and it is only through the reconnections, one after another after another, that today's humanity is born.

When we consider the small percentage of our genes that come from the Neandertals, we might be tempted to say that they contributed very little. But even the 2-3% of DNA that comes from Neandertals is significant. "These genes have provided our species with the ability to migrate to, and succeed in, numerous new environments, with different genes introgressed in different contexts, indicating that gene flow has been responsible for key aspects of human variation (different disease phenotypes, skin properties, etc.)" (Ackermann et al. 2015.) Add to this the strong likelihood that Neandertal interbreeding is just one page in a very long story of lineages diverging and converging. Even at the most basic biological level, we are what we are because we emerged by convergence and gene flow.

We are only beginning to discover how interbreeding makes us what we are biologically. A bigger question, one that requires more than science to answer, is how interbreeding affects the way we see ourselves. The old question was this: "what does it mean to be human?" In light of our new understanding of our evolution, our new question has become: "What does it mean to be an interbred, hybridized humanity?"

One thing it means is that we need to have a widened understanding of ourselves as a species seen historically. *Homo sapiens* is diverse today. But today's diversity pales in comparison to what came before. Were Neandertals really a different species? The very concept of species is in play. Pääbo, with his detailed knowledge of the breadth and complexity of past human genetic diversity, tends to think of our lineage not as a species but as a "metapopulation." He writes that "modern humans were part of what one could term a 'hominin metapopulation'—that is, a web of different hominin populations, including Neanderthals, Denisovans and other groups, who were linked by limited, but intermittent or even persistent, gene flow" (Pääbo 2015).

Rather than seeing our species as a static abstraction with clear definitions and natural boundaries, Pääbo is urging us to think of it as a dynamic, interactive, and changing web of interbreeding communities. A similar view is proposed by another team of experts. "We view the emergence of our lineage as a continuing dynamic (process) rather than an outcome (product); there is no clear starting point, or ending point, but rather an ongoing, repeating process of divergence and hybridization at multiple points in its evolutionary history. It is the dynamics of this repeated lineage divergence and remerger that has produced the variation observable in our genome…" Our modern *Homo sapiens* is not the result of "the directional accumulation of modernity." It is, instead, "a sporadic, flickering signal" (Ackermann et al. 2015).

When we look back over the past 200,000 years, we see many forms or regional variants of humanity. These include Neandertals, Denisovans, the "modern humans" from Africa, hybrids between them, and no doubt more local forms yet to be discovered. Whether we call them different species is mostly a matter of how we use words. Here we have stayed with traditional use and spoken of Neandertals as if they were a distinct species. And we have referred to the newcomers from East Africa as "modern." That is a convenient way of speaking, but it can be misleading in two ways. First, it implies that the "modern" newcomers were a new species. Second, it suggests that others were not modern but primitive or archaic.

Some experts suggest we rethink the use of words like Neandertal, not using them to refer to separate species but to local variants or sub-groups. "We would recommend that researchers consider everyone prior to living people who contributed directly to the variation in our lineage as human ancestors with regional population names like Denisovans and Europeans, rather than giving them species-level distinctions" (Ackermann et al. 2015).

More than terminology is in flux here. Our self-concept as a species is being transformed. Based on what we are learning about our origins, it is becoming clear that our evolution involves moments of expansion, travel, separation, divergence, and isolated adaptation. These moments of separation within sub-groups or regional lineages are offset by times of convergence and occasional interbreeding. Not only is there variation between groups, but also variation within them, most likely explained by localized divergence. Neandertals, for example, were a varied lot, and the ones that were first encountered by the newcomers from Africa were just one of many types.

What we see here is diversity within groups on top of diversity between groups. Diversity is kept in check by convergences that are frequent enough to keep the widespread human populations somewhat unified. This suggests three things:

First, the human genetic variation we see today goes deep into our past. It is shaped in part by regional interbreeding. At the same time, regional interbreeding is offset by migrations that cross the continents and re-connect populations from East Asia to South Africa to Western Europe. The result is that we are one species with variation that has been shuffled again and again.

Second, when we look back on our past, it is hard to see clear biological definitions or natural boundaries that mark off our form of humanity from others. Of course, it is true that living human beings are distinct from all previous forms for the simple and obvious reason that we alone are alive today. And it is true that all human beings who exist today are one species, according to any imaginable definition of that word. Looking back, however, the lack of clarity makes it hard to see how we can think of ourselves as distinct. We are not clearly different or distinct from Neandertals or Denisovans. Are they included with us in a wide circle of "human uniqueness," or excluded in order to preserve some force to the claim that we are unique? When we look back, distinctness and uniqueness seem to vanish in a grayish blur in which there are no visible lines.

Third, if oscillation between divergence and convergence is the way we evolved, then we should see divergence and convergence as creative and generative. By itself, divergence would lead to multiple separate human lineages. Convergence alone would reduce diversity and the overall viability of the human population. Together, the two are creative and generative.

Each of these three points leads directly to social and theological questions that will be explored in the final three chapters of this book.

8

When Adam and Eve Disappear

Recent polls suggest that Americans are divided almost evenly on whether Adam and Eve were the first humans. Those who accept this idea tend to think that the first human couple was created directly by God about 10,000 years ago. From that moment, we human beings were pretty much the way we are today.

Why do people believe this? Lack of education may play a role, but the real reasons often run much deeper than a simple lack of awareness. In fact, it can be insulting to suggest that people who stand up for a traditional Adam and Eve do not know any better. They are not being obstinate or willfully ignorant. They do not need to be told that their views conflict with science. They know that the truth of ideas is not based on familiarity or convenience. Some will say that rejecting a literal Adam and Eve means rejecting the Bible. Others are convinced that Adam and Eve provide the historical anchor that secures everything they hold dear about faith, family, and the purpose of human life. They find the old view socially reassuring and morally useful. Precisely because so many attach such great significance to the historical reality of a literal first couple, any discussion of the disappearance of Adam and Eve needs to be both respectful and honest. Deeply held values and convictions are at stake.

Today's science, on the other hand, is pretty clear about these things. We human beings are a mixed bunch, evolved over millions of years, and formed by diverging and converging. Is this upsetting? It does undercut old values and doctrines. Highly cherished religious and cultural ideas seem to have their foundations ripped from below. Compared to the old story of Adam and Eve, science seems to make it harder to answer the most basic questions about what it means to be human. Are human beings one? Are we unique or distinct among species in our status in nature or in our intrinsic value? Are we "fallen" or in need of redemption or salvation through grace?

Here in this chapter, we look at some of the ways in which the traditional story of Adam and Eve has played a role in answering these questions. Sometimes the answers were helpful, and sometimes not. By looking back, we can learn about mistakes to avoid. We will also come to a better idea of what we need to do to make sense of all the new scientific insight that is coming at us, at times so quickly that it is hard to understand, much less absorb.

How far is it possible to affirm the most important human values and religious convictions while fully embracing the scientific account of human origins? A good way to gain some perspective on that question is to look back over the past 500 years of recorded human history. How did religion and science interact with each other and with deeper, sometimes malevolent attitudes about human diversity and group superiority? The story is complex. But several features stand out clearly. Scientists and religious leaders alike held racist attitudes, and both were capable of twisting their convictions to support their prejudices. If there is a lesson to learn, it is that there is something in human beings that runs deeper than our ideas, scientific or religious.

A New Story, or More of the Same?

Today we know that our humanity emerged through a slow and complicated process of migration, modification, and mixing. Ancestral communities separated and went their own way, adapting biologically and culturally as they went. Their descendants, slightly changed during times of long separation, sometimes came back into contact with other groups. What happened next? For the most part, we can only imagine what these ancient encounters were like. Confusion, fear, violence, and acceptance were probably all part of the process. These emotions, these visceral responses, are still with us today when human groups encounter each other.

We can only imagine the ancient encounters. But when it comes to the encounters of the past 500 years, when great intercontinental migrations occurred, we can draw on historical records. How much might prehistoric human encounters of the distant past been like the ones we know about from recorded history? Were encounters with Neandertals and Denisovans anything like the encounters between Christopher Columbus and other Europeans when they came face to face with the natives of the Americas? Spanish and Portuguese sailors who reached the Americas around 1500 found themselves looking at a family of human beings from whom they had been separated for about 30,000 to 60,000 years, possibly longer but probably with some interaction. Both groups had changed over time, and some of the Europeans were not sure that the people they saw were really people. Europeans who entered sub-Saharan Africa at the same time were also confused by what they saw. The encounters that followed over the next 500 years were often filled with violence, disease, mass extermination, enslavement, forced relocation, and of course interbreeding. Some of the invaders may have exaggerated their confusion to justify their actions.

Does the story of human encounters over the past 500 years provide any clue to the ancient, prehistoric encounters? We will never know. We do know that some of the ancient encounters were preceded by as much as half a million years of separation. By contrast, the separation that came before the encounter between Columbus and the natives of the Americas was about one-tenth as long. We know what the Europeans thought of the peoples they encountered. We can only wonder what the Neandertals and the newcomers to Eurasia might have thought when they first saw each other.

The historical encounters over the past 500 years led to interbreeding. People from Europe, the Americas, and Africa, brought together by violence or desperation or choice, interbred and produced offspring. Thanks to technical advances, we know today that interbreeding also happened in the prehistoric encounters, when Neandertals met newcomers. The result did not always produce fertile offspring, but the mixing was successful enough to leave its legacy in our genes. As the technology of archaic DNA recovery advances, we are sure to learn much more about the details of the lineages that make up our past. How did groups move from place to place? To what extent did they coexist, keeping their distance except perhaps to trade with each other? To what extent did they merge with existing populations or replace them? As scientists continue their work, the secrets of the past are being opened up before us.

These techniques of genetic analysis are also being applied to the interbreeding that has occurred within the past 500 years, as well as to the millennia that came just before. In fact, the tools that are so useful in uncovering the distant past are now being marketed directly to consumers through online sites. Television programs such as *Finding Your Roots*, hosted by Henry Louis Gates, Jr., invite viewers to learn about the ancestry of celebrities. The most common finding is that any given individual's recent ancestors probably came from various continents. Discovering our personal ancestry through genetic analysis is becoming more common every day, mirroring on the individual level what is happening at the species level (Campbell et al. 2014).

Putting it all together, we are coming to see that the human beings alive today were created through a long history of complex interactions, as lineages separated and reunited. Whether this happened in the last few generations or a thousand generations ago, the overall pattern is repeated again and again. There was no clean line of descent that led to us as a species, just as there is no unmixed ancestry that led to me as an individual. Inside each cell in our bodies, our DNA witnesses to the multiplicity of our origins. The tangled web of our pedigree makes us what we are individually, as groups, and as one global humanity.

Science is only beginning to use these new tools of genomics and paleogenomics. New scientific insight is sure to come along at an ever-accelerating pace. Some of what we learn is sure to upset cherished notions about ethnic identities and native lands. And scientists will continue to do what scientists do. They will push against the horizons of our ignorance, answering questions once thought unanswerable. In just the past decade, research has presented us with surprising new insights, offering a new scientific account of our past. In the years ahead, scientists will continue their search for new archeological sites.

They will develop new techniques of analysis. The past decade has overturned many previous ideas, and the next decades will continue to do so, probably with even bigger surprises coming at an ever faster rate.

The new scientific picture is not the old polygenism, even though we now see how regional adaptations played a role in forming today's humanity. At the same time, the new picture is not some minor tweak on the strict "Out of Africa" story. Furthermore, the new picture does not support the idea that some 40,000 years ago, humans experienced some sort of cultural breakthrough, a cultural "big bang," marking a boundary between the pre-human and the fully human phase of our evolution. In every way, the new scientific story is more gradual, more diverse, more localized, and yet more global than what we thought just a decade or so ago.

The main challenge facing us is not scientific or technical. Science will go forward and do its work. Nor is the main challenge one of just trying to keep up with scientific developments, as hard as that is. Science writers will sort through the research and offer interpretations that are scientifically accurate and easily readable.

The real challenge is for all human beings alive today on this small planet to find a way to let more knowledge make us more human. It would be a tragic irony if all our new scientific insight into human origins only served to make us less humane. The risk is real because racism and ethnic conflict are very much alive today. The new information we are learning can be misused to exaggerate human diversity and argue for ethnic or genetic superiority. That has happened before, and it could happen again.

Why We Need to Learn from the Past

The story we tell ourselves about how we got here tells us who we are. Whether we base the story on the Bible or on the fossils, the full meaning of the story of our origins is found in the way we see ourselves and in the ways we treat each other. When we look back over the past 500 years, we see how theological and scientific views were both used to spin stories to justify racism, ethnic domination, and slavery. Throughout this period, some used the Bible to support racism. Others hoped that science would bring enlightenment and tolerance. Of course, there were people who argued for the end of slavery on the basis of scripture and of science. But far too many used these same sources to defend segregation, slavery, and mass extermination. If racism is our motive, we can use science or scripture to defend it.

However, if we wish now to turn our backs completely and categorically on all forms of racism, and if at the same time we seek to offer a view of one humanity rooted in theology and informed by science, it is instructive to look at the mistakes of the past. Doing so intensifies the urgency in our longing for a better future. And so we ask how Christians in past centuries responded to some of the social and intellectual upheavals of their times. In particular, how did they interact with the sciences as they tried to interpret and reinterpret the traditional story of Adam and Eve?

Looking back, we might think that the story of Adam and Eve would have undermined racist theories. The story, it would seem, should have provided direct support for three important arguments that should have slowed or stopped racist views. First, belief in a literal Adam and Eve tends to bolster the idea that humanity is one. We are all recent descendants of a literal, historical pair. Differences in culture and appearance notwithstanding, deep down we are all cousins. Second, the story of Adam and Eve gives us a lofty view of humanity as unique among species, created in the image of God and endowed with a human soul. Third, Adam and Eve provide the key element needed to make sense of the core claims of Christian theology. The sin of Adam and Eve was the grounding rationale for the need for salvation through Jesus Christ. On the surface, at least, these themes contradict racism. We are all one, we are all special, and we are all in the same predicament.

Here in this chapter, we look more closely at these three themes. In traditional theology, human unity, human souls, and human sinfulness are all tied to the view that Adam and Eve were literal, historical persons. Over the past 150 years, Christians have disagreed with each other on how to understand these three themes. Some drew on science, some on scripture, and some on both. All of them drew on their social location and on the social and economic values that were most important to them. How did they fit science and theology together, and how did they force both to fit a deeper agenda of racial superiority? Today we reject nearly all their conclusions. But we cannot escape the fact that like them, we are embedded in a dynamically changing context that is defined by scientific insight, theological ideas, political and economic forces, and cultural conflicts.

Racism and Human Unity

Debates about human origins are never just about origins, or so claims historian David Livingstone in his definitive study, *Adam's Ancestors*. He writes: "The question of human origins was not simply about science and species, it was about society and sex, cultural identity and racial purity. Bloodlines mattered—economically, politically, spiritually" (Livingstone 2008, 169). Implicitly or explicitly, the question of race was always present. Those with racist attitudes found ways to misuse science or theology, or a mix between the two, to serve the cause of white supremacy. Contrary to what some anti-religious/pro-science voices suggest, science did not "enlighten" people or free them from religious bigotry and racism. Despite what pro-Christian/anti-evolution voices might say, devout and God-fearing Christians held views that were utterly destructive of the humanity and dignity of others.

Between 1870 and the 1930s, the most widely held scientific view of human origins was polygenism. According to polygenism, today's human beings are divided into separate races that evolved independently in Africa, Asia, and Europe. Polygenists held that for perhaps a million years, each separate human group became the way it is today, and there was little if any contact between them. In their view, human differences run deep into the past. To strengthen their argument, they often exaggerated the measurements of present-day differences between these "races." The view came increasingly under attack in the 1930s and especially after World War II, partly because polygenism was used to justify claims not just of European superiority but of "Aryan" supremacy. Since the early 1960s, polygenism has been universally rejected by evolutionary biologists who are experts on human origins.

During the decades following 1870, some scientists rejected polygenism. The main alternative to polygenism was monogenism, which claimed that all human beings are one species with small and largely superficial differences. Charles Darwin himself rejected polygenism and endorsed a monogenic view. According to Darwin, "All the races agree in so many unimportant details of structure and in so many mental peculiarities, that these can be accounted for only by inheritance from a common progenitor; and a progenitor thus characterised would probably deserve to rank as man" (Darwin 1882, 608).

According to Darwin and the biologists who agreed with him on this point, humans evolved first as a unified population until they became pretty much like us, with bodies and minds like ours. Only then did they spread out to occupy different parts of the globe. Natural selection continued to work, and in time these groups were modified by their local environments. The critical question facing monogenists was how much change occurred after separation. Darwin himself was conflicted on this. Like nearly everyone in his time and place, he accepted the idea that Europeans are naturally superior to everyone else. He spoke of civilized and savage races.

Even though monogenists thought all humans were one species, some thought that local environments created differences almost as great as those found in various canine breeds, like bulldogs and poodles. Human differences were not just limited to things like skin pigmentation but affected cognitive ability and social capacities. Some humans, they believed, were so inferior to others that they benefited from colonization, even from slavery. In one of the sobering ironies from the history of science, polygenism and monogenism were both used to justify white supremacy and racism.

In the United States, many who embraced traditional Christian theology or Biblical literalism objected to evolutionary theories, polygenic and monogenic alike. Our theories of human origins, they insisted, must be based on the Bible. The distressing thing is that in the end, it really did not matter whether one accepted polygenism, monogenism, or a literal Adam and Eve. Each view was interpreted to support racism. In the case of the Bible, the challenge was to get beyond the fact that the opening chapters of Genesis plainly teach that all humans are recent and direct descendants of Adam and Eve.

If Genesis 1-2 presented a problem for supremacist views, Genesis 6-9 contained the answer. This passage contains the famous story of Noah's ark. Huddled together with the animals that took refuge in the ark, only eight humans survived the global disaster. These are Noah, his wife, their three sons, and their wives. After the flood, according to Genesis 9, Noah planted a vineyard, made wine, got drunk, and fell asleep naked in his tent. The youngest son, named Ham or "Canaan," came into the tent and saw Noah. Ham went out and told his brothers, who backed into the tent and covered their father without looking. According to Genesis 9:24-25, "When Noah awoke from his wine and knew what his youngest son had done to him, he said, 'Cursed be Canaan; a slave of slaves shall he be to his brothers.'" According to the interpretation, the curse endures forever. The biblical "Hamites" were the ancestors of Africans and other enslaved people, and their condition is mandated by God.

In a way, it was like a second fall. According to the standard interpretation of Genesis 3, all human beings are cursed by the first fall. But the descendants of Ham are doubly cursed, diminished even more in their loss of the original human capacities and suited only to serve others. In fact, the curse on Ham means that his descendants can benefit from slavery. Like domesticated animals that are bred to work and obey, they are not capable of self-control or self-determination. And yet they are not animals, but a lower form of humanity. This "adamic theology," as Livingstone calls it, sees interbreeding as possible and perhaps even beneficial to the descendants of Ham, but degrading to whites and destructive to white society.

The troubling thing is that it hardly mattered whether one turned to science or scripture or to a blend of both. Racist views of white supremacy could be justified many ways, and seemingly intelligent people felt morally justified in their attitudes by what they took to be sound arguments based on their highest authorities. Comparing two views, Livingstone writes: "Put simply, scientific anthropology bestialized slavery; adamic theology sanctified it" (Livingstone 2008, 182). Despite their agreement on the cash value of their views, the advocates of the two views criticized each other. Livingstone quotes Thomas Smyth (1808-1873), a Belfast-born Presbyterian preacher in Charleston, South Carolina, who complains about the scientific approach: "It would remove from both master and servant the strongest bonds by which they are united to each other in mutually beneficial relations….God is in this whole matter….The relation now providentially held by the white population of the South to the colored race is an ordinance of God, a form and condition of government permitted by Him" (quoted by Livingstone 2008, 182).

Perhaps Rev. Smyth did not fully grasp what was in the hearts of his hearers, some of whom not only owned slaves but may very well have been subjecting them to treatment cruel even by the standards of the time. Some may have been exploiting their female slaves sexually, then coming to worship with their white families to praise the God who had ordained their power while wondering what to do with unwanted pregnancies.

Christianity and Polygenism

One might think that science would have corrected Rev. Smyth's racism. Or perhaps that Christian scholars would have criticized scientific polygenism. That did not happen, at least not nearly enough. It is true that over time, Christians came to depend less on the curse of Ham in Genesis 9 and more on the unity of humanity in Genesis 1-2 for their views. Scientists came to recognize that polygenism went too far in exaggerating the differences that make up human diversity. These changes were worked out amidst the horrors of the American Civil War and World War II, both fought in large part over conflicting views of human worth and identity.

As much as anyone, it was the Nazis who ended polygenism by exposing its full implications. Scientific support for monogenism was starting to grow in the 1930s, but a compelling scientific case against polygenism was still a few decades away when notions of racial supremacy led to mass destruction and genocide. In the aftermath of World War II, polygenism was seen not just as a mistaken theory but a dangerous one at that. At least in the United States, however, post-war institutionalized racism remained, as incomprehensible and intolerable as it may have seemed to African American soldiers who fought and died to liberate oppressed peoples elsewhere.

At about the same time, the Catholic Church was struggling to take a position on evolution. The growing scientific support for the basic theory was hard to ignore. But polygenism was seen as a problem for Catholic theology. In 1950, Pope Pius XII released a Vatican encyclical that commented briefly on human origins. The encyclical was entitled *Humani generis*, and in it the Pope cracked open the door ever so slightly to biological evolution and even to the idea that humans have evolved. But Pius insisted on two limitations. The first has to do with the Catholic idea of the human soul. By its very nature, the human soul cannot come into existence by evolution. The question of the soul will be explored briefly in the next section. The second limitation, Pius writes, is that polygenism is unacceptable because it contradicts Catholic theology. According to the encyclical, "the faithful cannot embrace that opinion which maintains that either after Adam there existed on this earth true men who did not take their origin through natural generation from him as from the first parent of all, or that Adam represents a certain number of first parents" (Pius XII 1950, para 37).

That sentence is complex because it rejects two distinct versions of polygenism. Pope Pius is ruling out any view that suggests multiple human origins. This is the standard form of polygenism held by some scientists even as late as the date of the encyclical. But Pius is ruling out another view that he also sees as problematic. It is necessary, the document says, that a literal Adam and Eve, one human couple, be the historic parents of all human beings. No evolutionary biologist accepts this view.

Ruling out the scientific version of polygenism is helpful, but it is not clear why Pius objects to it. Is he concerned that the theory supports racist and supremacist ideas or that it undermines a doctrine of original sin? If the concern is racism, then offering a warning about the scientific version would be enough to make the point. But the Pope's main concern seems to be to defend the traditional doctrine of original sin. For that, a historical Adam and Eve are needed. "Adam" cannot symbolize some sort of founding population or community of humans. According to the encyclical, "it is in no way apparent how such an opinion can be reconciled with...original sin, which proceeds from a sin actually committed by an individual Adam and which, through generation, is passed on to all and is in everyone as his own" (Pius XII 1950, para 37).

At about the same time, Protestants were sharply divided on these issues. Many held to what they saw as the plain meaning of the Bible. When the scriptures speak on questions of human origins, the only question is how to discover the full historical meaning of the text. Others saw texts like Genesis not as historical or scientific accounts but as stories or sagas with rich symbolic meaning and enduring insight into the human condition. The gap grew between Protestants who insisted on a historical Adam and Eve and those who read the texts symbolically. The first group objected to evolution while the second saw no problem in letting scientists do their work, since it was largely irrelevant to theology. Neither group addressed such things as polygenism.

Here in this book, of course, we are rejecting all these alternatives. Science should be accepted as a valid source of insight and used to advance theology's understanding of humanity. It should not be rejected or ignored. The old polygenism, of course, is no longer advocated by any experts in human origins research. The new scientific picture is one of divergence, regional adaptation, and interbreeding. This is theologically significant, and the details of the new picture should be taken into account theologically. This task is especially urgent today, given the rapid pace of scientific advances in the study of human origins.

Polygenism and the support some found in it for racism are in the past, but they are hardly gone. As we discover more about the ways in which human ancestral populations migrated and then converged in regional contexts, some will be tempted to claim a new, scientific foundation for supremacist views. The old polygenism endorsed the idea that regionalized evolution on separate continents produced different levels of humanity. *The new view is not polygenism*, but it stresses the significance of regional or localized evolution and adaptation much more than the unmodified "Out of Africa" theory that held sway until about a decade ago.

It remains true that all human beings are descendants of the recent African community. But it is also true that global human diversity is shaped in part by encounters with older, regionally adapted communities. How far do these regional encounters explain human variation today? The scientific answer is "probably not very much." The popular cultural answer, however, might be something far different. The challenge facing all human beings today is to make sense of this science while refusing to let it play into notions of racial differences. To repeat: The new view is not the old polygenism but something altogether different. Even so, the new view can easily be twisted to serve the same interests of racist social and political attitudes that made polygenic views so appealing in the first place. The task for theology today is not to condemn outmoded polygenism but to do everything in its power to prevent the new view of human origins from being misused to support the old sad tale of racism and supremacy.

Human Souls

In Genesis 2, we read that God formed Adam from the dust of the ground and then breathed into the lifeless form the breath of life, and Adam became a living soul. Other animals have souls according to Genesis, but only human beings are animated by God's direct action, which awakens them to consciousness and gives them a capacity for a relationship with God. Often in the past, Christians connected this text with Genesis 1 and its statement that Adam and Eve were created in the "image of God." We who descend from Adam and Eve are endowed with this same status as creatures in God's own image, bearers of human souls. Many theologians such as John Calvin held that the image of God in humanity rests primarily in the human soul. Taken together, the question of the soul and of the image of God provide the basis for a grand view of humanity.

According to traditional theology, we humans are distinct among creatures in capacities, in our role in nature, and in our relationship to God. This distinction is ours from the beginning and is grounded in the way God created us. Human specialness rests in the claim that human beings are created in the image of God and that we alone possess a human soul. These two themes are closely tied together. In Chapter 9, we will look at what it means to say that humans are created in the image of God. Here, however, we look at the question of the human soul in the context of our origins. Our focus here is not so much on what it means to have a soul. Here we are asking how humans got souls in the first place.

We begin by looking once again at the Catholic position. In *Humani generis*, Pius is open to human evolution but not to polygenism or to the idea that evolution can produce a human soul. Recent Catholic statements agree. By its very nature, the human soul is not something that evolves. Instead, God creates it for each individual at conception. In the distant past, when human life had evolved to the point of readiness, God created the first human souls and joined them to adequately evolved human bodies. The point here is not biological but philosophical or ontological. For God to create human souls is not to violate evolution, because evolution cannot create human souls.

Speaking to the Pontifical Academy of Sciences in 1996, Pope John Paul said this: "With man, we find ourselves facing a different ontological order—an ontological leap, we could say." This does not violate evolution, John Paul insisted, because the human transition that comes with receiving a soul is not observable to the sciences. "The sciences of observation describe and measure, with ever greater precision, the many manifestations of life, and write them down along the time-line. The moment of passage into the spiritual realm is not something that can be observed in this way—although we can nevertheless discern, through experimental research, a series of very valuable signs of what is specifically human life" (John Paul II 1996, para 6).

John Paul does not say exactly what he means by the "very valuable signs" of the presence of the human soul, but perhaps he has in mind such things as early cave art or other evidences of human symbolic capacities. His point is that ancient art can be seen, but the inner transformation of humanity into rational, spiritual beings is something that cannot be seen, even if it happened right in front of us. The capacities are real, judging from the signs they leave. But the process itself can only be described "through philosophical reflection, while theology seeks to clarify the ultimate meaning of the Creator's designs" (John Paul II 1996, para 6).

According to Catholic teaching, the human soul must be created directly by God. This is not a matter of God intervening with created causes, for instance by triggering soul-building mutations in human DNA. It is a direct act of a different metaphysical order, outside the nexus of natural or created causes. In 2004, the Vatican released a document entitled *Communion and Stewardship*. The question of the origin of the human soul is informed by science, but it is directly addressed only by theology because it involves God's "non-disruptive" direct action. "With respect to the immediate creation of the human soul, Catholic theology affirms that particular actions of God bring about effects that transcend the capacity of created causes acting according to their natures."

The statement continues with these affirmations: "Catholic theology affirms that the emergence of the first members of the human species (whether as individuals or in populations) represents an event that is not susceptible of a purely natural explanation and which can appropriately be attributed to divine intervention." For this reason, "…it falls to theology to locate this account of the special creation of the human soul within the overarching plan of the triune God to share the communion of trinitarian life with human persons who are created out of nothing in the image and likeness of God" (International Theological Commission 2004, para 70).

The most recent Catholic statements do not insist on the existence of a literal, historical Adam and Eve. On this point, the recent documents differ from *Humani Generis*. The 2004 statement speaks of the first ensouled humans "as individuals or in populations." In a number of respects, many (but obviously not all) Protestants today agree in broad terms with the teachings found in recent Catholic thought. One difference, however, is that today's Protestants across the board tend to be more cautious than Catholics in speaking about the human soul. Recent Biblical scholarship has undercut the idea that the Bible supports the kind of soul/body dualism that the Vatican seems to be endorsing. Other Protestants, perhaps by turning to neuroscience or to philosophy for support, come to the same conclusion. Here the question is not whether Adam and Eve are historical but whether human creation involves two parts, the formation of the body and the creation of the soul. Even though earlier generations of theologians and Protestant confessional documents speak with confidence about the human soul, many Protestant scholars today agree that the concept lacks the biblical support and the conceptual clarity it once had.

When thinking about what happens after we die, attention has now shifted from belief in the immortality of the soul to the resurrection of the body. When thinking about human origins, Protestants today across the board do not embrace the idea that humans have a soul that is created separately from the formation of the body. Debate continues, of course, on whether God creates the human as body-and-soul through recent direct action or whether God creates the human through evolutionary processes. Either way, the action of God is a singular act. The human (as body-and-soul) is one reality, and God's work of creating us is one work.

Support for this unitary view can be found in early Christian theology. It is clearly taught in the writings of one of the preeminent theologians of the early Eastern Church, Gregory of Nyssa (335-394). In his treatise "On the Making of Man," Gregory offers two arguments that run counter to the Catholic view. First, he insists that while each human consists of body and soul, these two are unified as one. If so, they come into existence together, not first as body and then as soul (or vice versa). Second, he insists that God does not create us piecemeal, first in one dimension and then in another. That is demeaning, Gregory says, to the power and glory of the creator. If God made us one part at a time, "the power of Him that made us will be shown to be in some way imperfect, as not being completely sufficient for the whole task at once, but dividing the work, and busying itself with each of the halves in turn" (Gregory of Nyssa, *On the Making of Man*, 29.2).

Now of course, Gregory is writing here of the creation of each individual, not of the human species. And he is insisting that the whole thing is by direct divine action, not through evolution. Even so, the principle of the unity of God's creative act should hold. God creates humanity as a whole through biological evolution. The perfection of God the creator requires us to see God's work as one complete and unified work, not something divided between indirect evolutionary means and direct action.

What this means is that if we avoid soul/body dualism and hold to the unity of God's creative action through complex evolutionary processes, we come to see ourselves in our entirety as a product of evolution. We human beings, with all our subtle complexities and amazing capacities in all dimensions of our existence, come into existence as a species through evolution. The process is gradual, complex, and episodic. But at no point do we exist, completely evolved in body but as yet without a soul, awaiting divine intervention. There is no threshold moment when we cross a line between being ready for a soul and having a soul. We come into existence, as Gregory of Nyssa observed, not first in one dimension and then in another, but as a whole or a unity.

One benefit of letting go of the traditional idea of the human soul is that this concept should no longer be used to exaggerate the difference between humans and other animals. For too long, we have said to ourselves that other animals do not feel pain as we do or that they are not afraid of danger as we are because they do not have souls. Now we know that some animals, at least, form deep and enduring pair bonds. Some grieve the death of a family member or companion. As beneficial as it might be to value animals as species, it is even better to see them as individuals, the most complex of which are not so different from us.

Are Humans Fallen?

Other animals, especially those most like us, have complex moral capacities. Some have the ability to show what seems to us to be compassion or care. Some have the ability deceive others, to fight and even destroy rivals, and engage in conflict between groups. But when it comes to moral complexity, human beings seem to be capable of taking animal behavior to completely new and frightening levels. Only human beings seem capable of genocide, rape as a strategy of war, slavery, and the willful destruction of habitats and even of the planetary ecosystem. How do we explain this, and what role does it play in our theology?

A traditional Adam and Eve help to bolster the claim that humans are special. They also provide critical support for the idea that we alone are sinful and need salvation. In traditional Christian theology, especially the theology of the West, the human need for salvation functions as a critically important premise on which the whole of Christian doctrine is built. If we do not need salvation, we do not need a Savior. The fall of Adam and Eve explains why we need salvation. It is the necessary presupposition for classic western Christianity, shared by traditional Protestants and Catholics alike.

Of course it is possible to describe the human need for salvation without a literal Adam and Eve and without an event called "the fall." No one really disputes the reality of human sinfulness. For many, it is a simple fact so obvious that it requires no explanation. As individuals, becoming aware of our own moral failure is a part of growing up. For at least two centuries, dating back well before the time of Darwin, some Protestant theologians pointed out that the story of the Fall is really the story of us all. It is a tale of the loss of innocence, of breaking away from the security of home, disruptive and freeing at once. The story is the universal experience of coming of age. At the start of the story, Adam and Eve are fresh and youthful and innocent. Misguided or not, by their own choice they leave innocence for independence, and troubles quickly follow. Many Christian theologians have taken this story, set free from any claim to being literally or historically true, and built a richly meaningful reinterpretations of Christian theology around it, one focused on the breaking and restoring of relationships.

Not all Christians agree with that move. For some, a plain reading of the text leads directly to the notion that Adam and Eve are more than symbols. They really existed as the first people. To doubt this, many believe, is to doubt the truthfulness of the Bible and everything that it teaches. The scientific story of human origins conflicts with a plain-sense reading of the Bible. For that reason alone, evolutionary biology as a whole and the evolution of humanity in particular are seen as enemies of Biblical faith, at least for a significant number of Christians. This perspective, however, may be losing support among today's evangelical Protestants. Speaking to Christians who once held firm to a plain or literal meaning of the text, evangelically-affiliated scholars such as John H. Walton insist that "there may be a wider range of possibilities for a biblical and theological understanding of human origins than previously recognized. If it turns out (as I believe it does) that science offers evidence to the contrary, we are free to consider its claims" (Walton 2010, 204).

The reliability of the Bible, however, is not the only concern. Many Christians agree that texts like the opening chapters of Genesis can support a range of interpretations. But they still have a problem with deleting Adam and Eve from their theology. They see serious trouble here because of the way in which the broad doctrines of western Christianity seem to depend on the historical reality of Adam and Eve. Their focus is centered on Genesis 3 as interpreted in the Bible itself. The Epistle to the Romans seems to insist that events described in Genesis 3 really happened. In fact, these events *had to happen*. In Romans 5:19, for example, we find the tight connection Paul draws between Adam and Christ: "For as by one man's disobedience many were made sinners, so by one man's obedience many will be made righteous." If Christianity is to make any sense at all, Adam and Eve had to "fall" into sin—*actually* and not just symbolically.

Building on what Paul says in Romans 5, Christians in the West, Catholics and Protestants alike, have seen the fall of humanity into sinfulness as necessary if Christianity is to make sense. Their concern is not just the truth of the text. It is the coherence of the faith itself. After all, the key Biblical text of Genesis 3 does not even use the words "fall" or "sin." Taking its cue from the Apostle Paul, however, traditional western Christian theology has seen Genesis 3 as the necessary backdrop for the gospel. Later theologians such as Augustine (354-430) develop this theology even further.

Augustine is the preeminent theologian of western Christianity. He considered it impossible that a good God would make us sinful, with our propensities for deception, violence, and self-centeredness. But God did create Adam and Eve as free creatures, free enough to be sinless but also free to turn away from God and choose sin. In their freedom they chose disobedience and disorder. As a just response, God turns them out of the perfect paradise of the Garden of Eden, forcing them to live in a world that requires work if they are to survive. They wanted disorder, Augustine notes, and so God gives them what they want. Disorder now surrounds them in nature and disobedience destroys them from within. Because of the fall, Adam and Eve and all who descend from them cannot guide or regulate themselves.

This interpretation fits nicely within the wider frame of traditional Christian theology. It preserves the goodness of God as creator, protecting God from any blame for human evil and even for suffering in nature. More than that, however, it sets the stage for the central claim of Christianity about our need for Christ as redeemer. Only Christ can undo the damage caused by Adam and Eve. Christ's act of redemption reverses Adam and Eve's act of rebellion. Without a historic Adam and Eve, Christ has nothing to reverse. And in that case, Christ has nothing to do and nothing to offer. For some Christians, the matter is that simple. If there is no literal Adam and Eve, there is no need for redemption and no need for a Redeemer.

For some, it is almost as if we have to conclude that for Jesus to save us, we must first save Adam and Eve. Their concern is that the disappearance of Adam and Eve leaves a gaping hole in traditional theology. No wonder some refuse to let science knock out such a central pillar of their belief. Against all the evidence, they insist that by some definition, a historical pair named Adam and Eve not only sinned but transmitted the guilt and the brokenness of that sin to all living human beings, who are their direct descendants. Some simply reject the science in order to make this claim. Others look for more imaginative ways to try to reconcile traditional theology with science. Their goal is to preserve some sort of first parents, some version of a historic Adam and Eve, to serve as the critically important first pair, the origin of all human beings and our sinfulness.

As we already saw, evangelical scholars like John Walton are open to various readings of the text of Genesis. But they recognize that quite apart from the literal truth of the Bible, the logical coherence of traditional Christianity hangs in the balance. According to Walton, *"the historicity of Adam finds its primary significance in the discussion of the origins of sin rather than in the origins of humanity"* (Walton 2010, 203; italics in the original). The unspoken connector here, of course, is "no sin, no Savior."

We can see the connection when we think more about the traditional doctrine of sin. According to the traditional western Christian view, a valid doctrine of sin must make three claims. First, human sinfulness must be completely debilitating, so much that we cannot save ourselves no matter how hard we try. Far from making ourselves righteous, we just go on sinning. The grace of God in Christ is the only thing that can save us. Second, sin cannot be something that God creates. Everything God creates is good, and so sin has to come from the creation itself, as a free choice. Third, sin has to be universally present in all human beings, who are equally disordered or incapacitated by its presence. It is true, of course, that some human beings experience the destructive effects of sin far more than others do. In terms of sin's consequences, there is nothing equal at all about sin. For some, power or privilege keeps us largely immune from the bad actions of other. For many, however, the sinful actions of others destroy nearly everything. As true as that is, at the same time traditional Christian theology has claimed that all human beings are equally capable of sin. Or to put it more precisely, all are equally incapable of living holy or sinless lives. All are equally in need of salvation.

The simplest way to secure all three claims at once is through the classic theological interpretation of Adam and Eve. Their free act plunges all their offspring into a sinful condition that only grace can set right. Taken together, the three claims seem to make a literal, historical Adam and Eve a necessity. In the traditional view, we are all equally morally disordered and cannot save ourselves. God did not create us that way, but created us with the freedom to make ourselves that way. God alone can save us from this situation. For traditional theology, these truths can only be explained by a free human action that implicates us all. Only Christ can save, but only Adam and Eve can explain why we all need Christ. If we do not have Adam and Eve, we do not need Jesus. For these three claims to hold together in their traditional form, Adam and Eve must be actual historical people who literally were the parents of all humans.

But what is theologically necessary is scientifically unsupportable. Christians who accept the full traditional interpretation find themselves at an impasse. Do they remain Christian and reject a scientific view of human origins, or do they embrace science and surrender their traditional view of Christianity? Is there an alternative Christian view that respects the insights of science and yet does justice to the seriousness and the comprehensiveness of the Christian view?

Options for Christian Theology

The easiest way to avoid these problems while remaining Christian is to reject the traditional western connection between Christ and the sin of Adam and Eve. One way to do this is to look at Jesus Christ mostly as a good influence. Human sinfulness is passed along by bad examples. Christ reverses this by inspiring us to be perfectly loving and compassionate. When his moral and spiritual example catches on, people and institutions are transformed. Nearly everyone, Christian or not, recognizes at least a bit of truth in this perspective. At the heart of the traditional Christian message is the command that we are to imitate Christ. The idea that Jesus Christ is an exemplary human being who inspires people to do better is simple theologically. It fits easily with evolutionary theory because it has no need for a literal Adam and Eve. All it requires is that we see our own humanity as needing improvement. This theology is not threatened in the least by the loss of a historic or literal Adam and Eve. In fact, in very broad strokes, some Christians suggest that evolution provides a scientific basis for their notions of human uplift or progress.

But for many Christians, the task of improving ourselves to the point of imitating Christ is the very thing we cannot do. We need to be changed from our ordinary selves in order to become more compassionate and just. Christ must be for us something more than a good example. Christ must save us, not just by inspiring us, but by changing us. The idea that Christ saves us merely by providing a good example is naïve, especially in the face of magnitude and sheer horror of human evil. One good example from the past cannot meet the challenge of the serious trouble in which we find ourselves, facing both the end of humanity as we know it and the collapse of a hospitable planet due in large part to human action.

If science has undercut the traditional explanation of sin and salvation, and if alternative theories like Christ as moral exemplar do not work, what options are there for Christians today? Any credible option, it seems, must do at least two things. It must be existentially honest in facing the seriousness of our predicament, and it must be intellectually honest in drawing on the best insights about our origins. What is needed is a theology that takes science honestly and sin seriously. Christian theology today cannot be salvaged by minimizing the human predicament. But theology has nothing credible whatsoever to say about the salvation of humanity in Christ if it starts with a scientifically discredited view of humanity.

One way to try to move forward is to recognize the power of culture to shape humanity. In one respect, this is theology's oldest alternative solution to questions about the origins of sin, but now it comes with a new twist. It was Augustine's chief nemesis, the British monk Pelagius, who developed the view that cultural and environmental factors explain why we all seem to be morally disordered. Augustine is sharply critical of Pelagius. Over the centuries, most Western theologians have sided with Augustine, downplaying the impact of culture in making us sinful. It is our corrupted or fallen nature that warps our environments, not the other way around. Today, when we look at the question of culture and environment from the perspective of the long story of human origins, we see that our many ancestral environments have contributed again and again to the reshaping of our nature. The interaction goes both ways, of course. Evolving populations reshape environments. But environmental and cultural modifications rebound upon the biological forces that shape our species identity.

A particularly promising step in this direction has been proposed by David L. Wilcox, a population geneticist. He argues for a complexified environmental view that takes culture, natural environment, technology, and evolutionary biology all into account in the shaping of the human. Not just gene selection but gene expression or epigenetic variables are at play in making us what we are. All these processes "talk" to each other. According to Wilcox, "All of us as 'Adam's' cultural descendants are necessarily egoistic, with that impulse dominant over our altruistic impulses, in part, because the culture which nurtures and apprentices us determines the shape of the neural programming which makes us human." It is not too much of a stretch, Wilcox suggests, to see "fallenness" coming to form itself in us through these processes. "That cultural alteration likely also altered the selection pressures on epigenetic and genetic loci, increasing the power and malignancy of the fallen pattern."

On this basis, Wilcox sees in us an inherent sinfulness that is at once environmental and biological. He writes: "We are born as sinners because we can only become human by being nurtured by humans—who are all sinners. Adam's sin is and was therefore indeed our sin." The key to Wilcox's proposal is the way he links the environmental with the biological, not as nature *versus* nurture but something more like nature *through* nurture. All the way back, not to a historical couple called Adam and Eve but as far back as we can imagine our ancestry, there is something about it that Wilcox calls a "fallen pattern." On this he thinks theology and science agree: "We need a Savior!" (Wilcox, 2016).

One attraction in Wilcox's proposal is that he draws on recent scientific views of the multidimensionality of the evolutionary process. Today we recognize the dynamic interaction between populations, genes, gene expression, cultural artifacts including tools, and the surrounding environment. Wilcox suggests how this dynamic and interactive process we speak of so simply as "human origins" can, in fact, produce a creature of great intelligence, creativity, moral sensibility, and spiritual awareness. At the same time, according to Wilcox, this process has gone awry and followed a "fallen pattern." The achievement here is to preserve human fallenness without a historical fall. He does this to preserve our need to see Christ as Savior.

There is, however, another way to look at what Christ is doing. In the Bible and in the writings of some of the earliest Christian theologians, we find a Christ who saves us by *completing* humanity rather than *reversing* the fall. To the extent that this is true, our attention shifts away from the questions of sin and the fall. Our attention is turned instead to questions of human incompleteness. Then we see Christ saving us by bringing our evolved humanity to its fulfillment and destiny in eternal companionship with God. In Christ, all human beings are brought together and made one, and in this way it remains true that Christ saves us from sin. All these themes will be explored more fully in Chapter 10. In Chapter 9, however, we ask first about what it means to say that human beings are made in God's image.

9

Evolving in the Image of God

Genesis 1:26-27 says that on the sixth day of creation, after creating the heavens and the earth and all the other creatures, God made us. The text describes God speaking while acting: "Then God said, 'Let us make man in our image, after our likeness; and let them have dominion over the fish of the sea, and over the birds of the air, and over the cattle, and over all the earth, and over every creeping thing that creeps upon the earth.' So God created man in his own image, in the image of God he created him; male and female he created them."

The idea that we are created in the image of God is ennobling and uplifting. We may not know exactly what it means, but we know we like the idea. Over time, this little phrase has taken on a life of its own, exaggerated far beyond its significance in the Bible and probably well beyond its usefulness in theology. According to Jürgen Moltmann, theology gives a place of high honor to the "image of God," but the Bible does not. He writes that "the biblical traditions do not offer any justification for the central place given to this concept" (Moltmann 1985, 215). Robert Jenson, noting the out-sized role played by the concept in theology for two thousand years, comments this way: "One might wish some other notion had been given this comprehensive function, but it is too rooted in the tradition now to be displaced" (Jenson 1997, 53).

There is at least one good reason for keeping it. The idea that all human beings are bearers of the divine image is a powerful moral marker, a sanction against human violation. The moral significance of image-bearing is found first in Genesis 9:6, which makes the murder of the human being a violation of the divine. Even though the moral principle is often ignored, what would our world be like without this marker in place? This worry is captured in a haunting comment from the "Grand Inquisitor" passage of Fyodor Dostoevsky's novel, *The Brothers Karamazov*: "If there is no God, everything is permitted." Mass murder committed by regimes that officially embraced atheism makes the worry anything but hypothetical. In the past few centuries, the idea of God has largely disappeared from modern Western Europe. If God is absent, does being created in God's image mean anything?

To guard against the consequences of the idea that humanity without God lacks special moral status, post-war Europe and indeed the whole world has endorsed the notion of human dignity. For example, the Charter of the United Nations affirms the "dignity and worth of the human person." The Universal Declaration of Human Rights, adopted by the UN General Assembly in 1948, begins by declaring that the "recognition of the inherent dignity and of the equal and inalienable rights of all members of the human family is the foundation of freedom, justice and peace in the world." These words may be ignored in practice, but no one seriously denies that the world is somewhat better because they are universally affirmed.

The idea of human dignity goes back to the Roman orator Cicero, and so some of the advocates of the secular concept of human dignity claim that it is not a religious idea but based in a more universal rationality or natural law. But the idea of dignity is nurtured in the Christendom of the pre-modern West. Whether the secular idea of dignity remains conceptually coherent and compelling when stripped of its theological roots is a matter of debate. Wolfhart Pannenberg argues that the connection between human inviolability and its Biblical roots in the *imago Dei* is more than just a historical tie. Speaking of the secular idea of human dignity, Pannenberg warns against detaching "the idea of universal inviolability from its basis in the Bible." If we do that, our widely-proclaimed human dignity will not have "any solid basis at all" (Pannenberg 2001, 177).

If Pannenberg's warning comes anywhere near the truth, and if we value a world in which something like human dignity is universally affirmed, we cannot help but see the value and significance of the idea that all human beings bear the divine image. But is the idea of humanity created in the image of God even comprehensible when we come to see humanity as evolved? The idea may have moral value, but does it have intellectual integrity? Our challenge is to try to understand this idea for our time, making sense of it by redefining it in light of the science of human origins.

Challenges Old and New

What, then, does it mean to say that humans are made in the image of God? Exactly how do humans reflect or mirror the divine? This is a theological question, but many have claimed that it is possible to point to something in us that is the visible mark of the image. Recently it has been suggested that the theological notion of image of God is connected with the scientific question of human uniqueness. After all, only humans are in God's image, so there is an implicit claim of uniqueness embedded in the theological idea. So we must ask: Is the image something we can point to empirically, perhaps with the help of science? And if so, is it possible to use recent studies in human origins to point out when and how evolving humans became uniquely human? Is there a moment when we evolved into the image of God?

Part of the problem in answering these questions is the sheer multiplicity of views on what is meant by the biblical and theological idea, "image of God." In the first section of this chapter, we try to sort this out. But sorting out the various meanings is not the same as settling on the one right meaning. In fact, I suggest that there is no one right meaning. In that respect, "image of God" really is a troublesome idea, not so difficult that it should be jettisoned, but not easily defined or neatly fit into a larger theological view.

Nearly every theological definition of the image of God points to something about us that marks the presence of the image. Does the "image of God" have a marker that we can see? The challenge here is that the *imago Dei* is first of all a theological concept. Does the theological idea have an empirical correlate? Is the image of God definable in a way that makes it observable? If not, how do we know when it is present in humans or absent in other creatures? Most theologians have accepted the idea that the image of God has empirical correlates, even while disagreeing on what the correlates might be.

The question of the markers of the image set the stage for an even bigger challenge, the one generated by the scientific study of human origins. Can we identify the empirical correlates of the image with enough clarity so that when we look back over the story of human origins, we can point to concrete evidence that shows when and how the image evolved? The final sections of this chapter turn to that question. Some have suggested that there is a close tie between the theological idea of the image of God and the scientific or empirical question of human uniqueness. Is human uniqueness empirically definable and scientifically defensible? If so, when did the key features that comprise human uniqueness first appear? What signs survive even now to provide evidence of its appearance? For example, perhaps humans are unique because of the rich, symbolic cultures we create. It has been suggested that the first signs of symbolic cultures are the empirical correlates of the theological concept of the image of God. If so, then science can help us discover when symbolic culture first appeared. The scientific support for these claims will be examined more closely in the final sections of this chapter.

Before we come to these questions, however, we turn first to the challenge of trying to define theologically what is meant by the idea itself. The claim that we are created in God's image is a theological idea. According to Moltmann, "The human being's likeness to God is a theological term before it becomes an anthropological one. It first of all says something about the God who creates his image for himself, and who enters into a particular relationship with that image, before it says anything about the human being who is created in this form" (Moltmann 1985, 220). What does *imago Dei* mean in Christian theology? What does it say about human beings?

Defining the Image

Over the centuries, Christians have defined the idea of the image of God in various ways. Here in this section, we will look at some of these options. Our list identifies eight different approaches. Any list like this is arbitrary, and most lists have fewer items than ours because they cluster some of our options together. When individual theologians actually go about developing their own view, they almost always combine elements from several approaches into a more complex package. The point here in teasing the approaches apart is to clarify whether these definitions of the image of God point to something that can be observed or measured. The overarching question in this chapter is whether the image of God has empirical correlates that can be seen as emerging in the evolutionary process. To get at this question, we start by distinguishing eight different approaches to the question of what it means to be in the image of God.

The first approach is one of the oldest and most common in the history of Christianity. It is often called the "substantive approach." That label is being used here in a strict sense for the claim that the image of God in humanity is located primarily or exclusively in the human soul as a substance distinct from the body. The soul may function in close harmony with the body, but as a substance it is ontologically distinct from the body as a different kind of reality. All animals including humans have bodies, but only human beings have human souls. In Chapter 8, we saw how recent Vatican statements endorse this view. Human bodies evolve, but the human soul is made of something that cannot evolve. It is directly created by God and joined to human bodies when God sees that evolution has reached the point when body and soul can function together. At the same time, of course, the Vatican teaches that the whole person is made in the image of God, and this means that we are created for communion with God, a theme developed in another one of our approaches. But for Catholic teaching, it is the ontologically distinct human soul that makes communion with God possible.

The second approach is to say that the image of God refers to a set of capacities that human beings ordinarily have. This approach is called the "capacitative approach" or the "structuralist approach." Chief among these special God-like human capacities is rationality. Recent theology has expanded this list to include a richer view of human life that pays attention to human values and emotions. Some who adopt a capacitative approach combine it with the substantive approach. In their view, special human capacities are based in the soul. In that case, human capacities did not evolve but were directly created by God, as the Vatican teaches. Soul/body dualism, however, is widely out of fashion today, and it is possible to hold to a nondualist capacitative approach. In this view, special human capacities are based in the human brain and are made possible by the brain's own stunning complexity. Anyone who accepts evolution accepts the idea that the human brain evolved. If so, then at some point in human evolution, the brain becomes able to support the capacities that make us special. A challenge to the capacitative approach is sometimes embedded in the question: How should we think of individual humans who lack one or more of the defining capacities? We know we can lose our capacities over time, but we insist that we are still included in the community of those who bear God's image because of biological and social ties. A similar question might ask: What are we to think of those ancient human ancestors who did not quite meet the standard of fully human capacities but who are tied to us biologically?

A third approach starts by turning in a completely different direction. It points to the special functions of human beings, and so it is often called the "functional approach." The focus here is not on human capacities but on our assigned role in the created order. This approach can claim perhaps the strongest biblical support. Immediately following upon Genesis 1:26-27, we read: "So God created man in his own image, in the image of God he created him; male and female he created them. And God blessed them, and God said to them, 'Be fruitful and multiply, and fill the earth and subdue it; and have dominion over the fish of the sea and over the birds of the air and over every living thing that moves upon the earth.'" Those who advocate the functional approach point to the way this text leads directly to the command to fill the earth and have dominion over it. When God creates other animals, they are not given any command. Human beings are in the image of God and commanded to exercise dominion, or care of the household, over the rest of creation. That divinely mandated function is what it means to be in the image of God. Many wonder, however, what capacities humans need if they are going to perform this function.

The fourth approach points to human social relationships and family ties. If God is triune, existing as persons in relation, then humans reflect God by being persons in relationship. This "relational approach" claims biblical support by pointing to the phrase, "male and female he created them." It is mainly in relation with others that we experience and enjoy what it means to be in God's image. The great strength of this view lies in pointing out how human personhood is shaped socially. Culture and language all depend on social interaction. In our evolutionary past, the patterns of human social relations interacted dynamically with our emerging capacities for language, technology, and art. One challenge faced by this view is that many other creatures live in close relationships, but that does not mean they also share God's image. Another is how to think about individuals who live alone.

A fifth approach puts the focus on the relationship that humans have with God. To mirror or image God is to live in relationship with God, a relationship that shapes everything about us. One way or another, this "God-relationship approach" is found in nearly every theological explanation of the meaning of the image of God. Some theologians, however, favor this approach because it seems to avoid some of the problems associated with other approaches. What counts in the end, they say, is not our capacities or our human relationships, as important as these may be. The key issue is that God addresses us and invites our response. According to Robert Jenson, what sets us apart from other highly sophisticated creatures "is that we are the ones addressed by God's moral word and so enabled to respond—that we are called to *pray*." We are "praying animals" (Jenson 1997, 58-59; italics in original).

The sixth approach, best called the "Christological approach," is based on an interesting problem that faces any Christian theologian interpreting the biblical meaning of the image of God. Genesis 1 speaks of the first male and female as created in the divine image, but several New Testament texts speak of Christ as the only one who is truly or originally in the image of God. For example, in Col 1:15 we read that "He [Christ] is the image of the invisible God, the first-born of all creation..." The rest of us are invited by God "...to be conformed to the image of his Son" (Rom 8:29; see also 2 Cor 4.4 and Heb 1.3). In early Christianity more than today and in Eastern Christianity more than in the West, these texts figured prominently in an interpretation of the meaning of the image of God as Christological. Underlying the interpretation is the doctrine of the Trinity. Christ is fully divine, one with God and yet distinct in personal identity, a complete and perfect reflection of God. Strictly speaking, only Christ is in God's image. We who are creatures are called to live in the image of the Image, as ancient writers often said.

In some ways, only the Christological approach can claim to take into account the relevant Christian scriptures. In addition, it alone makes sense of Christ as the very expression of God who at the same time brings us into relationship with God. On the other hand, with the exception of contemporary Orthodox Christian theologians, this view is largely ignored in recent theological definitions of the image of God. To its credit, the Vatican's *Communion and Stewardship* clearly lifts up the centrality of the Christological approach. We read that "what it means to be created in the *imago Dei* is only fully revealed to us in the *imago Christi*. In him, we find the total receptivity to the Father which should characterize our own existence, the openness to the other in an attitude of service which should characterize our relations with our brothers and sisters in Christ, and the mercy and love for others which Christ, as the image of the Father, displays for us" (International Theological Commission 2004, para 53).

The seventh approach can be called the "developmental approach," and it is associated with the early theologian Irenaeus (130-202). He thinks of Adam and Eve not as God's final version of humanity but more like children or adolescents, immature forms of what is to come. According to Irenaeus, "it was necessary that man should in the first instance be created; and having been created, should receive growth; and having received growth, should be strengthened; and having been strengthened, should abound; and having abounded, should recover [from the disease of sin]; and having recovered, should be glorified; and being glorified, should see his Lord" (Irenaeus of Lyon, *Adversus Haereses* 4.38.3). This is God's preferred way to create us, Irenaeus says, and creaturely humility is our proper response. He advises: "You ought first to keep within the bounds of humankind and from there partake in the glory of God. For you do not make God, rather it is God who makes you. If then you are the work of God, await the hand of your fashioner who does all things at the due time, the due time for you, that is, who is being created. Offer him a soft and pliable heart and retain the shape which your fashioner gave you" (Irenaeus 4.39.2). What is most interesting is that for Irenaeus, what it means to be in the image of God is only realized at the end of the process. Adam and Eve and all of us are fully human, even though none of us yet has reached maturity in God's image. What it means to be in the image of God is stretched out conceptually by Irenaeus. According to God's plan, it takes on different moments or shapes as we move along toward maturity in Christ. The full meaning of humanity is hidden in Christ in the end. In our present form, we still have a long way to go.

Our eighth and final approach is perhaps the most difficult and yet the most provocative. It can be called the "undefined approach" or perhaps the "apophatic approach." Its central claim is that to image the undefinable God means that we humans are undefinable. The classical statement of the undefined approach is found in Gregory of Nyssa. God exceeds our comprehension. If we are in God's image, we might think we can understand God if we can understand ourselves. But in fact at our very core, in our own mind, we discover that we cannot comprehend ourselves. The human being, therefore, bears "an accurate resemblance to the superior nature, figuring by its own unknowableness the incomprehensible Nature" (Gregory of Nyssa, *On the Making of Man* 11.4). What makes this most remarkable is that in Gregory's view, it is theologically necessary that we cannot truly know ourselves. If we image God, we are incomprehensible.

This view is rare among theologians, but we do see it in Karl Rahner. He defines the human as "an indefinability come to consciousness of itself" (Rahner 1966a, 107). Rahner quickly adds that there is much about us that "can be defined, at least to some extent." But just as God is indefinable, the core nature or identity of the creature who is in God's image is also indefinable. We find a hint of a connection between this view and the Christological approach in 1 John 3:2. We read that "it does not yet appear what we shall be, but we know that when he appears we shall be like him, for we shall see him as he is." Another text that points in this direction is 1 Cor 13.12: "For now we see in a mirror dimly, but then face to face. Now I know in part; then I shall understand fully, even as I have been fully understood." The use of "mirror" hints at image. Together these texts suggest that the full meaning of being in the image of God is not yet clear to us, but in Christ it shall be made clear.

One of the more famous statements about humanity's lack of definition comes from the Italian Renaissance and the writings of Pico della Mirandola. According to Pico, God gave all the nonhuman creatures their specific forms or natures, but the human is given an "indeterminate form" (Pico, para. 4). At the beginning of our creation, our form and therefore our future are left undefined so that we humans may define ourselves.

It is of course just a coincidence but striking nonetheless to compare Pico's "indeterminate form" with a comment found in a recent scientific article that describes the oldest fossil (LD 350-1) that seem to belong to the genus *Homo*. The researchers argue that this small fragment, this first trace of humanity, a partial jawbone and a few teeth, seems to belong to *Homo*. But then they add that this first sign of humanity is undefined in form or species. Their closing sentence reads this way: "For the present, pending further discoveries, we assign LD 350-1 to *Homo* species indeterminate" (e et al. 2015). It is as if the science of human origins echoes in its own way Pico's "indeterminate form" or Rahner's "indefinability."

Refining the Image

This list of eight possible approaches to the image of God is best seen as a palate of colors from which a fully developed view might draw. A fully developed concept of the image of God might pick up any of these approaches. They are almost always found in combination, not in isolation. Several of the approaches seem to depend logically on one or more of the others. For instance, is it possible to say that we are created for relationship with God without claiming that we have the capacity for such a relationship? Or can a Christian even imagine being in relationship with God without Christ at the core of the connection?

One version of the "God-relationship approach" tries to tamp down any claim for special human capacities. Recent writings by Joshua Moritz provide the best example of this approach. Moritz wants to avoid making the claim that humans are unique among species or special in any theologically decisive way, as if we alone have the capacities for a relationship with God. One reason for doing this is to see that other creatures have advanced capacities, too. Another is to see that our close human relatives, such as the Neandertals, had the capacity to be in relationship with God. Our kind is not biologically or culturally unique, but we are nevertheless the ones chosen by God for relationship. The key theological idea here is "election," God's gracious choice based completely on God's favor and not on our capacities. We are in the image of God and enjoy a relationship with God solely because God chooses us. Protestants familiar with the classic doctrine of election will see how Moritz's view stands in that tradition.

It is not that Moritz denies that capacities are important. His view is criticized by Smedt and Cruz, who claim the obvious when they says that "relationships cannot be conceived of without positing at least some capacities to engage in them. If humans do not have any capacity to respond to God, they do not have a relationship with him, but more something like attachment" (Smedt and Cruz 2014). Moritz's counterpoint is that capacities alone are not enough. So theology does not need to identify and defend a set of capacities as uniquely human. Many species have what it takes for a relationship with God, but only humans are elected to be God's image-bearers. "The image and likeness of God in humans is not essentially recognized by reference to any capacities, qualities, skills, behaviors, or even souls (*nephesh*) that *Homo sapiens* might possess in distinction from animals or other nonhuman creatures." The image of God in humanity rests completely in God's "free historical action" (Moritz 2016, 204).

Like Moritz, Robert Jenson is also cautious about linking the image of God to a unique set of capacities. Of course it is true that a relationship with God requires capacities on our part. But are these capacities uniquely human? Jenson asks: "Wherein do we resemble God that other creatures do not?" It is not clear, Jenson says, that we can really point to some capacity or "complex of qualities, supposedly possessed by us and not by other creatures." But we are in God's image, and the foundation of this theological truth lies in God's action and not in our uniqueness. According to Jenson, we are special among all the creatures because God speaks to us. When that happens, God puts us in relationship to God.

This view is based in part on a strong notion of God's free choice, a view rooted in Protestant notions of salvation by grace alone. Just as the redeemed are saved by grace alone, so humanity is chosen to bear God's image by grace alone, according to this view. Other theologians, however, are not convinced. Our complex capacities as creatures matter theologically. The image of God points to something special about us that can be defined and observed. For example, some say that the image of God rests in special capacities that humans typically exhibit at a level higher than other animals.

A better alternative draws on elements of the developmental, Christological, and apophatic approaches. From the developmental approach, we see that the image of God is not static or definable as something that already exists. The advantage is that it does not tie down the meaning of God's image to any particular moment or form in human history. Instead, it tends to see the whole developmental process, from our origins to our destiny, as a dynamic and transformative movement toward the full realization of the divine image. The key here is that the *whole movement in all its forms* is in the image of God. To this we can join the Christological approach, which adds a decidedly different dimension with its claim that being in God's image means being open to redefinition in Christ. And finally, the apophatic or undefined view insists that nothing we can point to unlocks the core meaning or the central definition of our humanity.

How we understand the image of God will determine our answer to the question of its empirical indicators. What are the markers of the image? When did they first appear? The science of human origins challenges theology today to address these questions. One way to answer these questions is to define the image of God in terms of distinct capacities, and then to ask whether we can find signs in the past that these capacities have arisen. The main attraction of this approach is that it combines theology with serious attention to the science of human origins. In the final sections of this chapter, we explore this possibility and ask critical questions about its scientific support. Our conclusion is that despite its attractions, this approach lacks the kind of clarity and support that is needed to make theological sense of what it means to be human.

On the other hand, if we think of the image of God as humanity developing in Christ into a final state beyond comprehension, the question of the evolution of image-markers is freed from the task of trying to find a clear transition point between the ancestors who bear God's image and those who do not. The developmental, Christological, and apophatic approach sees the whole hominin evolutionary emergence as a movement toward the image of God. Within this whole, it is not necessary to define before or after or who is in and who is not. In fact, by a kind of reverse logic, the apophatic approach looks for the absence of empirical markers. In doing so, it does not look for gaps in our knowledge that science may someday fill. Instead, the apophatic approach finds its empirical support in the ways in which our knowledge leads to unanswerable questions about what it means to be human.

At this point it is worth quoting 1 John 3:2 once again. In simple terms, this text gets to the heart of the developmental, Christological, apophatic view being advocated here. The writer reminds us that "it does not yet appear what we shall be, but we know that when he [Christ] appears we shall be like him, for we shall see him as he is." We do not know what we are, much less what we shall be. But we know that the meaning of our humanity, now hidden from view, is somehow being defined for us all in Christ. This theme will be developed more fully in Chapter 10. In the final sections of Chapter 9, however, we will look more closely at the alternative approach. If the image of God is defined in terms of human capacities, held by human beings at a unique level in terms of consciousness and creativity, when did these capacities first evolve? Are there markers in the distant past that show us when and where they emerged?

The Image of God and Human Uniqueness

Are we human beings unique among creatures in our level of cultural creativity and sophistication? Any reasonable comparison among species suggests that this is so. No other form of life today is able to do what we do at our level. When we think just of our technological capabilities, and when we consider our powers to transform the natural world and perhaps even to change our own species intentionally into something *transhuman*, it becomes obvious that we are one of a kind in nature.

The reality of unique human cultural sophistication seems obvious. The question, however, is whether theology should define the image of God in terms of human uniqueness. For most of us, it seems intuitively obvious that there is at least some connection between the two ideas. But as we saw earlier, several of the most important ideas that cluster around the theological notion of the image of God tend to drop away when we identify the image of God mainly or exclusively with our advanced cultural capacities. The relational dimensions, especially the God-relationship as defined in the Christological approach, fade from view when we link the image of God only to human sophistication. And at the same time, looking only at our sophistication and power can blind us to our capacity for evil. Too often, the sophistication of some means the domination of others. There is also the risk that we will miss the truth contained in the undefined or apophatic approach. How well do we really know ourselves? Do we know ourselves well enough to trust ourselves with the sophisticated technologies we are developing? A complete account of what it means to be human cannot limit its focus to our cultural sophistication.

There is another, more immediate problem to be faced when we connect the image of God with human cultural sophistication. How can we fit this claim together with the scientific story of human origins? Of course it is true that our capacities for culture have evolved. Today, we stand out as unique among living species, and evolution tells us the story of how this came about. But what we see in the scientific story of human origins is that there is no defining transition point when humans became unique. If we humans are unique as a result of evolution, when did we become unique? Seen through the scientific and historical evidence, the evidence of uniqueness becomes elusive. There is no clean or sharp line between pre-unique human ancestors and fully unique present day humans. We can find no distinct threshold event, no identifiable moment of transition. To be clear, we are not disputing the fact that humans today are amazingly different from other creatures. What we are challenging is the claim that there is a clear and defining moment in the past when uniqueness happened.

If there were such a moment, it must be relatively rapid and completely transformative. The key moment of change would of course build upon slow changes that came before. But in itself, the transition point must be a breakthrough moment, somehow taking past evolution to a whole new level. Could this have happened in the distant human past, some 40,000 years ago? Some argue that this is indeed the case. They point to claims made by some biologists that the movement of evolution can be "punctuated." Evolution is usually slow, but its pace can be "punctuated" with periods of rapid change. David Wilcox modifies this idea in order to suggest that after a long evolutionary process, human ancestors develop sophisticated cultures rather suddenly. Wilcox takes up a suggestion made by John Walton, who notes that the Genesis text calls attention to the transition from human biology to human function.

According to Wilcox, the notion of being made functional helps us "understand a 'punctuated' model of gradual human creation. Even if the genetic substratum is 'prepared,' it does not automatically produce a functionally modern brain." Wilcox goes on to suggest that there was indeed some decisive moment in the past. "A point of sudden appearance of the image might have been produced by the impact of a threshold event in cultural transmission. This could happen due to the profoundly culturally driven (re)shaping of the cerebrum which takes place during early development. It is, after all, those culturally driven qualities which make humanity unique. Such a transformation would not necessarily leave a detectable physical trail in the form of transformed skulls or altered genetic loci. But would such an event take an extended theophany, or perhaps a miracle of neural transformation to make Adam (or a group) truly unique, to jump the gap to full humanness?" (Wilcox 2016).

Wilcox is probably right that it *could* have happened this way. But did it? One possibility, of course, is that there were many evolutionary and cultural breakthrough moments that added new functions over time. In this view, the key transitions happened not all at once in some grand way but many times in small ways. But that is not what Wilcox has in mind. Is he right in saying that there was a moment of dramatic change in the past, a transformative "sudden appearance" that by itself brought a group of ancestral humans across a theologically significant dividing line?

While disagreeing in several ways with Walton and Wilcox, J. Wentzel van Huyssteen seems to agree on the idea of a rapid and profound advance in cultural performance. He speaks of "a remarkable evolutionary shift in human capacities, or rather a shift in performance." He adds that "all evidence points to the fact that behaviorally modern humans were astonishingly quick in developing their creative artistic skills over the shortest period of time" (van Huyssteen 2006, 168, 171). Van Huyssteen's work, especially his 2006 *Alone in the World? Human Uniqueness in Science and Theology*, is a breakthrough theological treatment of the question of the evolution of the image of God in humanity. He is clearly correct in claiming that evolution is the source of what theology identifies as the image of God. But does the image of God evolve suddenly or slowly? Does it arrive in a burst of cultural creativity or in many small steps, spread widely in time and space?

How Does a Spiritual Animal Evolve?

To support the claim that it happened quickly, van Huyssteen draws on some ideas advanced by experts in human origins research. The evidence, he suggests, points to the sudden emergence of human minds. Cultural artifacts from 30,000 to 40,000 years ago provide tantalizing hints of the mental and perhaps even the spiritual complexities of the people who made them. Best known and most convincing among these artifacts is the early cave art of France and Spain. According to van Huyssteen: "What we learn from the sciences of paleoanthropology and archeology, then, is that human uniqueness emerged as a highly contextualized and physically embodied notion, where the symbolizing minds of our distant ancestors are stunningly revealed today in the materiality of prehistoric cave art. The emergence of these symbolizing minds, revealed in the cave art, is also linked to the unique ability of spoken language, and a remarkable cognitive fluidity as expressed in the ability to generate mental symbols, to think, to feel, to reason, to imagine, and to plan." As van Huyssteen sees it, these developments are tied "directly to the emergence of religious awareness" (van Huyssteen 2006, 212).

Even if we accept the idea that evolution is "punctuated," it is hard to imagine any single transformative event with such sweeping consequences. The most recent evidence, in fact, tends to undermine the idea that it happened suddenly, as Wilcox and van Huyssteen suggest. Almost as if he were sensing that the tide of scientific opinion might shift, Robert Jenson is cautious about putting too much stock in one interpretation offered by scientists. Jenson warns: "Theology need not share the anxious effort to stipulate morphological marks that distinguish prehumans from humans in the evolutionary succession." Even more pointedly, he says: "Theology need not join debates about whether, for example, the cave paintings were attempts to control the hunt or were thanksgivings for the hunt, were 'magic' or 'religion'" (Jenson 1997, 59-60).

But does Jenson go too far with his warning? It seems that where van Huyssteen connects theology and science closely, running the risk that science over time will shift in its interpretation, perhaps Jenson separates them too cleanly, going overboard in keeping them apart. What does it mean for theology to "join debates"? Scientific debates must be argued based on expert interpretations of the most recent evidence. Theologians bring no special competence to that task. If that is what Jenson means, then of course he is right. But theology does have a stake in science, especially here in the science of human origins. Theology cannot be "un-joined" from these debates. The sad tradition of theological disinterest in science must be rejected as having no place in the future of theology. Van Huyssteen is surely on the right track in connecting the theological question of humanity with the scientific question of human origins.

When theology reflects in depth on the science of human origins, it comes to a new theological understanding of what it means to be human. "The theological notion of the *imago Dei* is powerfully revisioned *as emerging from nature itself*," as van Huyssteen puts it (van Huyssteen 2006, 322). The human capacity for a relationship with God is rooted in human biology and shaped by all the processes at play in our origin. "Our very human ability to respond religiously to ultimate questions, through various forms of worship and prayer, is deeply embedded in our species' capacity for symbolic, imaginative behavior, and in the embodied minds that make such behavior possible" (van Huyssteen 2006, 267). An evolved creature, as van Huyssteen claims correctly, bears the image of God.

The question is when and how this happened. Did it happen suddenly or gradually? Did we become fully human in a flash, or was it a slow, step-by-step process? In the decade of the 1990s and for a few years following, many leading experts in paleoanthropology endorsed the idea of a sudden explosion of culture. Some used phrases like "cultural big bang" to reinforce what they saw as the explosive speed and force that characterized the cultural changes that occurred somewhere around 40,000 years ago. They pointed to cultural artifacts found especially in Europe as tangible signs of a rapidly advancing human symbolic culture. Drawing on the classic form of the Out of Africa theory, they suggested that all human beings today descend only from a new form of humanity. When these newcomers faced novel challenges in Eurasia, they quickly adapted and advanced to a new level of cultural sophistication in order to meet the demands of their environment. Culturally modern humanity was the result.

This view accepts the idea that the bodies and brains of these first culturally modern humans were not much different from what they had been over the previous 100,000 years. In many ways, these humans resemble other animals. But something new and unique was emerging, something different from all other animals and all earlier humans. Van Huyssteen puts it this way: "However much humans share with primate or hominid culture, we cannot avoid acknowledging the emergence of distinctive cultural properties that highlight both the evolutionary continuities and discontinuities in the cultural behavior of *Homo sapiens*, earlier humans, and other primates. And the most distinctive discontinuity is the emergence of language and the symbolic capacity of the human mind. It is language that engages the interactive minds of the social group, and that enables the social world beyond an individual's own lifetime to be defined symbolically" (van Huyssteen 2006, 226).

Then comes the key theological claim. The rapid cultural advance thought to have occurred beginning about 40,000 years ago is linked to the theological concept of the image of God. "It is this astonishing dimension of human cultural behavior that is unique to modern humans and that suggests the origins of a spiritual sense. A sense of the ineffable, the sacred, the spiritual is part and parcel of how human beings have coped with their personal and social universe" (van Huyssteen 2006, 226). Two important claims are made here. The first is that cultural advance has something to do with the image of God. The second is that both emerged rapidly. Unique cultural sophistication and unique theological status as creatures in God's image emerged together recently.

Our concern here is not with the claim that humans have a spiritual consciousness, or that this has evolved, or even that we might link this spiritual consciousness with the image of God. The question is whether it evolved suddenly or slowly. It is obvious that it emerged, but *when* and *how*? Did we become culturally sophisticated moral and spiritual beings in a flash? Is there really a clear dividing line between *before* and *after*? Yes, according to van Huyssteen. He speaks of the "recency" of the cultural "big bang, the Upper Paleolithic revolution and the explosive growth of human creativity around 45,000 years ago" (van Huyssteen 2006, 66). Then he claims that "all evidence points to the fact that behaviorally modern humans were astonishingly quick in developing their creative artistic skills over the shortest period of time" (van Huyssteen 2006, 171).

It is not that van Huyssteen does not look at counter-evidence. He cites experts who warn against giving too much attention to the period around 40,000 years ago in Western Europe. Look more widely, they argue, and further back in time. Van Huyssteen notes their warnings "against a profound Eurocentric bias and a failure to appreciate the depth and breadth of the African archeological record" (van Huyssteen 2006, 177). Before these advances appear in Europe, a number of advanced technologies appear first in Africa, sometimes tens of thousands of years before "modern" humans reach Europe. These findings, van Huyssteen writes, should serve as a warning: "For those of us who still want to talk of the amazing 'creative explosion' of the western European Upper Paleolithic, and thus of the suddenness with which this new 'package' of consciousness appeared in western Europe and the comparative speed with which it replaced the old Neandertal way of life, it is very important to remember that in Africa and Asia, we find important precursors of the 'creative explosion.' It is certainly in Africa that we must seek the earliest evidence for what we might later want to call the 'human explosion,' and theologians who are engaged in interdisciplinary dialogue with paleoanthropologists have to be careful not to prolong a Eurocentric bias" (van Huyssteen 2006, 177-178). He even recognizes that "it follows naturally that the symbolic activities we see in Upper Paleolithic western Europe could have occurred before *Homo sapiens* communities reached present-day France and the Iberian Peninsula" (van Huyssteen 2006, 179). But in the end, van Huyssteen sides with the argument for sudden cultural advance in Europe as the decisive evolutionary basis for human uniqueness.

In the past decade, however, confidence in the whole idea of a sudden cultural explosion has eroded. Mounting evidence suggests that we need a new and more complex view of how human beings become culturally advanced. Experts in the field mostly now defend the view that cultural sophistication arose, not in a flash in Europe, but over time and in widely separated places. The new view is based in part on new findings and even more on the technical advances that allow scientists today to reassess dates associated with previous discoveries. And precisely because the story of the pathway to becoming human is critically important to theology's answer to what it means to be human, our final task in this chapter is to summarize the new view and then to ask what it means for our theology of the human.

New Paths, New Meanings

Did human cultural sophistication burst on the scene some 40,000 years ago? In the past decade or so, we have learned a number of things that cast doubt on this view. At the very least, we know now that the appearance of human cultural sophistication is a more complicated story than we once thought.

Among other things, we have come to see the Neandertals in a new light. We know that about 40,000 years ago, when Neandertal numbers seem to be in decline and the anatomically modern humans are making their way across Asia and into Europe, Neandertal culture was roughly equal in sophistication with anything else going on anywhere. As we saw in Chapters 6 and 7, Neandertals probably made personal decorations from feathers and eagle talons. They made what appears to be the earliest cave engravings in Europe, simple by later standards but the oldest known residential markings. These date to just under 40,000 years ago. In 2016, we learned that the very time when anatomically modern humans were only first starting to appear in East Africa, Neandertals created a cave structure in southern France. Dating to 176,000 years ago, this structure is among the oldest surviving human structures. And perhaps most surprising of all, we now know that Neandertals interbred in multiple encounters with the modern humans. The enduring identifiable traces of their independent genetic legacy is found in the DNA in our bodies. All this should give us pause when we point to the modern human advances in Europe 40,000 years ago as something that was suddenly human.

Of course, the paintings that appear on the caves of Spain and France, most especially at Chauvet-Pont d'Arc and at Lascaux, are truly dramatic expressions of the creative human spirit. But even here, the story is complex and the artists mysterious. The paintings at Lascaux date to a little more than 17,000 years ago. At Chauvet, on the other hand, human occupation of the cave system is divided into two periods. Each period stretches for thousands of years. The earliest paintings date to about 37,000 years ago. They youngest ones are about 28,000 years ago (Quiles et al. 2016). Taken together, the creation of the cave art from Chauvet to Lascaux spans 20,000 years. During that time, people came and went.

In fact, armed with the latest techniques, researchers can compare the DNA of these first modern Europeans with later individuals. When they look back at the human populations of Europe roughly from 45,000 to 15,000 years ago, the pattern that is revealed is one of turnover with some admixture. The key point is that these new findings seem to suggest that the people who painted the great cave art of western Europe around 35,000 years ago have disappeared (Fu et al. 2016). The Europeans of today are largely descended from later invaders. In other words, the Upper Paleolithic human ancestors who painted Chauvet have vanished almost without a trace. Modern Europeans are descended in large part from invading waves of people that came from the Middle East, starting about 14,500 years ago. In that case, it becomes harder than ever to claim that these first artists were key to the process by which we all became suddenly human.

Perhaps most intriguing of all is the discovery that paintings made in caves on the southern tip of the Indonesian island of Sulawesi are as old as the earliest in Europe, probably even slightly older (Aubert et al. 2014). Their simultaneity creates a real mystery for today's scientists. If art was invented twice (or more) at roughly the same time, what triggered its invention? If it was carried from one place to another, how did it spread so quickly without leaving at least some trace in between. Or might it have been invented, at least in part, in creative cultures of Africa many tens of thousands of years before it was taken to Eurasia?

Right now, these are only questions. Future findings might shed light on the details of the origins of human cultural sophistication. At this point, the evidence seems to line up against the idea that there was one sudden flash of cultural inventiveness. And yet it is true: human cultural sophistication evolved through the interplay of biological and cultural processes, if not in one big flash, then perhaps in many smaller breakthroughs. Theologically, the point is that when we look back over the story of how this might have happened, we see no bright lines. There are no distinct threshold events that mark the temporal or cultural boundary between humans in the image of God and those who are not quite there yet. All this sets the stage for Chapter 10. By erasing the old lines between the human and the non-human, science itself invites theology to think all over again about what it means to be human. For this, Christian theology looks to the figure of Jesus Christ.

10

Christ Makes Us All One

Hidden away in John Calvin's commentary on the New Testament Letter to the Galatians are some remarkable words about how Christ transcends all ethnic differences. According to Calvin, in Christ "there is neither Jew nor Greek. The meaning is, that there is no distinction of persons here, and therefore it is of no consequence to what nation or condition any one may belong." Why is this so? "Because Christ makes them all one. Whatever may have been their former differences, Christ alone is able to unite them all." Then, as if he is standing before his neighbors and urging them to welcome the refugees who flooded Geneva in the mid-1500s, Calvin writes this: "Ye are one: the distinction is now removed (Calvin, *Commentary on Galatians* 3:28).

Calvin's key phrase, "Christ makes them all one," is the overarching theme of this final chapter. Our aim here is to draw on the latest insight from the science of human origins in order to offer a constructive theological interpretation that is decidedly Christian and yet refuses to compromise when it comes to accepting the validity of science as a source of insight.

Of course, science being science, we expect new findings to be reported at any time, prompting experts to challenge current interpretations and offer new insights. Theology that engages science must try to find a way to keep up with new findings. But more than that, theology that engages science must be willing to revise its own ideas in light of current, well-founded insights. I believe this can be done in a way that is completely faithful to the core of the Christian theological tradition. In fact, if anything, my conviction is that the truth of Christianity and its practical value in today's world are only strengthened by the science of human origins. Seen this way, science is theology's best friend, not always easy to get along with but always enlightening to talk to.

Not everyone, not even all Christians, will agree with what this chapter has to say. Christians who hold to the literal historical truth of the Bible will object. Those who believe that a historical Adam and Eve are necessary to explain the origins of human sinfulness and the human need for redemption will also balk at what they read here. The non-religious will see this whole book as a fool's errand, a misguided enterprise for deluded readers. And of course, all who follow religious paths other than Christianity may justifiably be offended by the theological claims made here.

The purpose of the claims made here about Christ and humanity is not to pretend to have final answers or to resolve old questions. Our main purpose is to call Christians to humility and repentance. To follow Christ is to renounce all racism, self-privileging, and inhospitable exclusion. All I can do at this point in the chapter is to challenge critics to read on and to be open to the possibility that the latest science can be brought together with a reinterpretation of the core convictions of the Christian tradition, and that the result is something urgently needed in today's world.

Concerning the idea of an historical Adam and Eve, our claim is simply that Christian theology has nothing to fear from their disappearance. The end of Adam and Eve as historical figures is a gift to Christianity. They have never been the main point of Christian theology, and for theology today to cling to them is itself a dead end. Their disappearance invites Christians to think more deeply and clearly than ever before about the true and living End for all humanity. For Christians, the true End of Adam and Eve, of all humans past and present and future, is Jesus Christ, understood as all humanity brought together into relationship with God. That Jesus Christ once lived in Roman-occupied Palestine is, of course, critically important to what we mean by Christ. But from the very beginning, Christians confessed that Christ was much more than one individual human being. Fully human, Christ is in a profound way *all humanity*, one with all in order to transform or redeem all. He is the bringing of all humanity to its destiny of fellowship with God. And in doing that, Christ is bringing all humanity together with itself. Put another way, Christ brings us together in order to bring us to God. Or as Calvin puts it, "Christ makes them all one."

Looking back over the millions of years of human evolution, we find a story of repeated divergences and convergences. With the help of scientific analysis, we look back as far as we can see. Stretching out in distant times before us, we see *Homo erectus*. Before that, there was *Australopithecus*. And even earlier, *Ardipithecus*, only barely seen at all. These are our predecessors in all their variant forms, many still unknown, some living before or after others, some living at the same time in distinct forms overlapping for hundreds of thousands of years. Groups expanded, separated, moved, modified their environment and were modified by it, and then reconnected. As a flowing stream is separated by an island and then rejoins, human communities separated, changed, and then came back together genetically. To ask which group is human is to ask which branch is the stream.

Diverging, we became many things. We became Denisovans, Neandertals, and many other unknown ancestral groups. By being them, we became us. Converging, we are still on our way to becoming one thing, one unified global human community. For Christians, our hope for one unified global human community has a concrete form. It has a name. It is Jesus Christ. Faced by all the racism and xenophobia that still work their violence and evil in our world, Christians defiantly and faithfully proclaim that Christ makes us all one.

Christ Completes Humanity

Who, then, is this Jesus Christ, the true and living End of all humanity? In the sections of this final chapter, we try to explore briefly what this might mean by focusing on four closely related themes. Christ is the one who completes, unifies, defines, and fulfils humanity.

The idea that the creation of humanity finds its completion in Christ is hardly new. In 1 Corinthians 15:45, Paul refers to Christ as "the last

Adam." In other texts, Paul contrasts Adam and Christ, always with the idea that Christ completes Adam by bringing humanity to its full and final state. Over the centuries, theologians have spoken of Christ as "the second Adam." In the 1830s, for example, Friedrich Schleiermacher wrote this: "As everything which has been brought into human life through Christ is presented as a new creation, so Christ Himself is the Second Adam, the beginner and originator of this more perfect human life, or the completion of the creation of man" (Schleiermacher, para 89). For Schleiermacher, God's work of the creation of humanity is not finished until humanity is completed in Christ.

As the one who completes creation, Christ is already present in the beginning as the destiny toward which creation moves. Schleiermacher puts it this way: "For although at the first creation of the human race only the imperfect state of human nature was manifested, yet eternally the appearance of the Redeemer was already involved in that" (Schleiermacher, para 89). A century and a half later, Wolfhart Pannenberg notes that "our destiny as creatures is brought to fulfillment by Jesus Christ" (Pannenberg 2001, 210). And in the recent Vatican document, *Communion and Stewardship*, we read: "Just as man's beginnings are to be found in Christ, so is his finality. Human beings are oriented to the kingdom of Christ as to an absolute future, the consummation of human existence" (International Theological Commission 2004, para 54).

But it is Irenaeus of Lyon, writing in the late second century, who provides the most richly developed interpretation of what it means to say that humanity is made complete in Christ. For Irenaeus, humanity at the beginning is created in an incomplete form. Adam and Eve are youthful creatures, almost childlike. God intends for them to grow up, but they have other plans. When the time is right, Christ gets God's project back on track by opening once again the pathway for human communion with God and most of all for the full flowering of human maturity.

To make his point, Irenaeus has to refute the idea that a perfect God can only make things that are perfect or complete from the beginning. Irenaeus contrasts God and the creation. God is perfect, but creatures need growth. God's glory is most fully expressed in humanity living and growing toward completion in Christ. A long sentence from Irenaeus that was already quoted in Chapter 8 is worth repeating here: "Now it was necessary that man should in the first instance be created; and having been created, should receive growth; and having received growth, should be strengthened; and having been strengthened, should abound; and having abounded, should recover [from the disease of sin]; and having recovered, should be glorified; and being glorified, should see his Lord" (Irenaeus of Lyon, *Adversus Haereses* 4.38.3). In our own time, Pannenberg reiterates the point made 1800 years earlier by Irenaeus. "In the story of the human race, then, the image of God was not achieved fully at the outset" (Pannenberg 2001, 217).

It is abundantly clear that Irenaeus is not thinking here about evolution. He had no clue that humanity has evolved. For him, the changes are developmental, not evolutionary. We must not lift Irenaeus out of his context or force him into ours. And yet his ability to see change as part of God's plan has inspired thinkers from his time until now. In fact, precisely because of the theological challenge posed by the science of human origins, with all its changing forms as the central feature of its story, Irenaeus is read and paraphrased frequently today. We can hear the distant echo of Irenaeus in these sentences by Pannenberg: "The way of creation to humans. . .takes the concrete shape of a succession of forms. Each of these is called into existence as an independent creature. None is simply a means to human existence. Not every line in the sequence leads to humanity, but as a whole they all form a basis for its emergence. The multiplicity of forms expresses the inexhaustible wealth of God's creative power" (Pannenberg 2001, 115). In these sentences, Pannenberg goes far beyond Irenaeus and his developmental idea. Here, Pannenberg extends the idea of developmental change and embraces evolutionary change as God's way of bringing today's humanity into existence.

We know today that humanity has evolved. Can we interpret the long process of human evolution as God's creation? For theologians all the way from Irenaeus to Pannenberg, the long, drawn-out process of human evolution can be seen as God's creative act because it is moving toward its destiny in Christ. Step by step, humans are being made complete in Christ for communion with God. It is theology's task to try to describe this. It the human task to live it in its fullest reality. As Pannenberg writes: "If they do not join in creation's praise of God, humans fail to be what they ought to be according to their destiny, namely, settings and mediators for creation's fellowship with God" (Pannenberg 2001, 115). More than just thinking about it theologically, becoming complete and fulfilling our destiny in Christ is something we do even as Christ does it in us.

When Christians in the past thought about completion of humanity in Christ, they thought mainly about how it applies to each individual. For the individual, what does it mean to be made complete in Christ? At its core, it means eternal fellowship or communion of humanity with God. It also means the transformation of the individual. Over the centuries, Christians have used many metaphors and images to try to visualize what this transformation will be. Sprouting seeds, hatching eggs, and butterflies emerging from cocoons have all served as illustrations. Those looking for greater eloquence and sophistication might turn to Dante's *Paradise*. In its elegant simplicity, 1 John 3:2 says that what we shall be is not knowable now. In time, it will be revealed in Christ. As Irenaeus recognized, the full and complete meaning of the image of God is yet to be realized in each of us. We will each know it when we become it, and we will only become it when we see it in Christ.

Our focus here, however, is not the transformation of the individual. Our concern is the human species, or more precisely the entire human ancestral web of species, the whole seven-million-year process of the hominin emergence. In fact, we want to claim here that excessive focus on individual transformation is not a particularly Christian idea. If Christ makes us all one, it follows that we cannot have Christ privately. We agree here with Pannenberg, who says that "humanity as a whole, and not just this or that individual, is destined for fellowship with God" (Pannenberg 2001, 224).

In order to complete our humanity as a whole, Christ must also unify it and make all its disparate elements into a whole. For Paul and Calvin, Christ unifies humanity by bringing all living human beings together in one new humanity, preserving and yet transcending our distinct ethnic and national identities. This theme will be explored in the next section. In exploring this theme, however, I will add to it in two key ways. First, I will ask how the science of human origins expands what we mean by the unifying work of Christ, complicating the diversity but making the work of unification all the more important. Second, drawing upon traditional theological ideas while also going beyond them, I believe that Christ unifies all humans alive today with all in the past and all who are yet to come.

Christ Unifies Humanity

Christ unifies humanity, first of all, by transcending all the barriers of culture, color, and class identity, making us into one diversified yet unified humanity. We can think of this as horizontal unification, stretching out as it does across the entire breadth of humanity today. To believe in Jesus Christ is to embrace the truth of this unity as a gift while seeking to live into its full reality.

Christ also unifies humanity, secondly, by making us one with all who came before and all who are yet to come. Think of this as vertical unity, reaching deeper into the past than we can see and farther into the future than we can imagine. We who live today are remarkably different from those who came before, and those who come long after us may be different from us to an even greater degree. It is whole sweep of human emergence that is taken up by Christ into companionship with God.

When we think first about how Christ unifies living human beings by transcending our national and cultural differences, we can draw on a tradition of texts and ideas that go back to the Bible itself. Paul writes that "you are all one in Christ Jesus" (Gal 3:28), and this theme is echoed elsewhere. Recently, biblical scholars have looked at these texts through the lens of the experience of those who, by free movement or forced migration, have been part of the modern relocation of people from one part of the globe to another. Sze-kar Wan, for example, draws on new ideas about human hybridity. He suggests that Paul sees Christ unifying us not "by erasing ethnic and cultural differences but by *combining these differences into a hybrid existence*" (Wan 2000, 126; italics in the original). Wan suggests that this text does not cancel ethnic identities. It includes every identity with all its distinctive features. Where the original text says that there is "neither Jew nor Greek," the underlying message of the gospel is best expressed when we paraphrase the original to read: "You are both Jew and Greek, both free and slave, both male and female, for you are all one in Christ Jesus" (Wan 2000, 127). In the unifying work of Christ, ethnic identity is preserved and yet unity is achieved. According to Wan, "what I think Paul is calling for in Galatians is for each cultural entity to give up its claims to power" but not "its cultural specificities" (Wan 2000, 126).

But where in the long history of ethnic encounters do we find any evidence that anyone willingly gives up a claim to power? What we see is domination, economic and sexual exploitation, and sometimes extermination, often in the name of Jesus Christ. Perhaps as nowhere else, here it is that the Christian who looks to Christ in faith yearns for forgiveness and for release from the patterns of brutality, domination, and unrestrained sinfulness of past human interactions. What is more evil than slavery, rape on a mass scale or as a military tactic, or genocide? To proclaim the gospel of Jesus Christ is to denounce completely and utterly all of these actions and the racist and supremacist and self-privileging notions on which they depend. But it is also to hold to the hope for the redemptive power of Christ, doing in us and among us what we cannot do for ourselves. In spite of sin, Christ makes us one.

A visionary book that leads us in this direction is Brian Bantum's 2010 *Redeeming Mulatto: A Theology of Race and Christian Hybridity*. He narrates the long and painful social history, especially in America over the past four centuries, that results in hybridity and the birth of children of "mixed race" identity, or the "mulatto." Bantum writes: "This mulattic rhythm of personhood is a mark of humanity universally, while humanity's history is an unfolding drama of refusals and transgressions" (Bantum 2010, 96). By rehearsing this history, Bantum invites Americans to be honest as a nation about who we are.

But more than that, Bantum offers a message of hope. If there is redemption, it comes to us through the "Redeeming Mulatto," who is Jesus Christ. As Bantum puts it, "We must see this mulatto Jesus as not only for a particular people, but we must begin to see how the mulatto Jesus stands before us to remake all people" (Bantum 2010, 111-112). Precisely by being like each of us in all respects (as Christian tradition has long held), Christ is in a position to redeem us. By his human hybridity or mulattic existence, Jesus Christ unifies in his own body the humanity so often painfully divided.

We find an odd and slightly puzzling signal of the hybrid or mulattic identity of Jesus in the opening verses of Matthew. For reasons that are not exactly clear, the gospel writer tells us about the ancestry of Jesus. The genealogy of Jesus is traced back through male descent to Abraham. In addition to Mary, the mother of Jesus Christ, four women are also mentioned. That is itself is unusual when compared to other genealogies from the same time and setting. But what makes these four women stand out is their disruptive identities. Rahab, for instance, is described in earlier texts as a prostitute living in the ancient city of Jericho. Two Hebrew spies enter the city in preparation for a military assault. She hides them and then becomes one of them, eventually becoming an ancestral mother of Jesus. While there is some debate among scholars, there is strong evidence to suggest that all four women are ethnic outsiders. The men are all Hebrews going back to Abraham, but the women disrupt the pattern and mix up the bloodlines. The original author of Matthew's gospel and the first Christians readers would have seen this immediately.

Why are these four women included at all? "One explanation," Richard Bauckham writes, "is that the four women were understood to be Gentiles, and were included in order to show that the Messiah, whose male ancestors in his direct descent from Abraham could not, by definition, be Gentiles, nevertheless had Gentile ancestors, thereby suggesting his suitability to be the Messiah for Gentiles as well as for Jews" (Bauckham 1995, 313). Biblical scholars continue to debate this interpretation. But the idea that Christ unifies by bringing diverse ancestry together in his own body is arresting, at the very least. It fits directly with Bantum's "Redeeming Mulatto." Reframed in the context of the science of human origins, with its patterns of divergence and convergence, Christ is convergence in its most concrete and redemptive form.

The Convergent Christ

If today we know anything through science about the humanity of Jesus, we can say with a high degree of certainty that he, like us today, had Neandertal ancestors whose DNA he carried in his human genome. The gospel writer Matthew makes a point about the Gentile women among the ancestors of Jesus. Following the example of Matthew, theology today should take note of Jesus's Neandertal ancestors. And of course, it is not just the Neandertals that are included here, but all of the ancestral forms of humanity that come together to make us what we are and to make him like us.

Where Matthew was concerned to show the genealogical connection between Jesus and Abraham, a similar genealogy in Luke's Gospel traces Jesus's ancestors back to Adam and Eve. For us, of course, Adam and Eve have disappeared. When we speak of them at all, we mean something like human ancestry all the way back, at least to the divergence of the hominin line from the line that led to chimpanzees. What, then, are we to do with these texts? We might ignore them on the grounds that their references to Adam and Eve make them meaningless in light of the science of human origins. Or we might find in them a challenge for theology in our own time. Jesus Christ had ancestors. They matter theologically. And now we know just how mixed and varied they were, how far back they stretched, and how diverse they were in form and behavior.

The challenge for Christian theology today is to connect what we know about human origins with what we believe about Jesus Christ. Scripture itself connects Christ with Adam and Eve. Our task is to recognize his connection with all humanity in all our present diversity and in all our past and future forms. The Christ of Christian faith unites us horizontally, bringing us together in one diverse yet unified new humanity. And at the same time, Christ unifies us vertically. Like us in all respects, his ancestral roots lie deep in the evolutionary past. In one functional and bodily unity, his genome is the convergence of evolved sequence variations carried generation to generation from our human and pre-human past.

Suggestions along these lines are found in Pannenberg's writings. He claims that the Logos or Christ "is concretely united with the plurality of creatures by humanity, or more strictly by the one man who for his part integrates humanity into a unity as himself the 'new man.'" By integrating all the diverse forms of ancestral humanity, Christ makes one new humanity. At the same time, Christ joins this one humanity with all other creatures. As Pannenberg puts it, Christ "is the integrating center of the world's historical order, which is grounded in the Logos and will find its perfect form only in the eschatological future of the world's consummation and transformation into the kingdom of God in his creation" (Pannenberg 2001, 64).

In light of today's science of human origins, the time has come for theology to build upon and extend the theological claims of classic Christianity. The early church wrestled with the question of humanity and divinity in Jesus Christ. They insisted that Christ is at once fully human and fully divine, uniting humanity and divinity so as to bring all humanity into the fullest possible relationship with God. In their time, the central question was how to think about the unity of divinity and humanity in Christ. They insisted that the humanity of Christ is full and complete. For us today, the new, critical issue is not the unity of humanity with divinity, but the unity of humanity with itself. Ancient creeds confess that Christ is the union of divinity and humanity. In that unity, humanity and divinity remain what they are, and yet are made one. To this we simply add that in Christ, humanity with all its distinctions and divisions is itself being made one, remaining what it is in all its diverse forms, and yet unified.

When ancient theologians said that Christ was fully human, they were clear about Christ being completely human, lacking nothing of our humanity. They thought, of course, that human beings are all essentially the same. One instance of humanity is like any other. Now we know that humanity has changed profoundly over time, and that it will continue to change. In which form of humanity is Jesus Christ? Just as the early church confessed that Christ is fully or completely human, we can only make sense of this historic confession if we add that Christ is not merely one form of humanity but humanity in all its forms.

In the writings of Karl Rahner, we find hints that take us in this direction. He writes: "Christianity radicalizes this existing and known unity of mankind with its teaching that all human beings, despite differences of race and history, diversity of cultures, are not only called into existence by one and the same God, but also have one and the same final destiny, which consists in attaining God's self-communication in itself." For Rahner, God's "self-communication" is

Jesus Christ. As God's self-communication, Christ comes into reality in every human in every evolved form and context. According to Rahner, "Christian faith is aware of a universal history of salvation, common to all mankind, existing from the very outset, always effective, universally present as the most radical element of the unity of mankind…" (Rahner 1981, 160). Rahner's phrase, "the most radical element of the unity" of humanity, encapsulates what we mean by "humanity in its entirety." The key to understanding what this might mean is to see the unity, not as something inherent in the creation, but as the gift of the Creator.

If we try to say what this means in the language of divergence and convergence, we can say theologically that God creates through divergence in order to gather all things together in the convergent Christ. Rahner uses terms like "salvific will" to describe God's action and "universalism" to describe its result. He writes: "The universalism of the one salvific will of God in regard to all mankind, which establishes the final unity of mankind, is the sustaining ground of all particular history of salvation and religion." Here, Rahner is clearly suggesting that in order to save us, God must also unite us, making one final unity out of all our diverse forms.

In a recent comment on the unexpected discovery of a previously unknown population of *Homo erectus*, the journal *Nature* provided an editorial entitled "Humanity's Forgotten Family." The discovery was a few teeth and a part of a jaw dating to 700,000 years ago in Indonesia. After commenting that the odds against such a discovery are "ineffably remote," the editors go on to say this: "It is possible that many human species once existed, but became extinct with such finality that even those few that were fossilized have since disappeared, leaving absolutely no trace that generations of a distinct species lived and died on this planet—a kind of double extinction, without hope of memorial or discovery" (*Nature* 2016).

These words, unusually poignant for a science journal, highlight just how far-reaching our claim is that Christ unites all humanity, horizontally and vertically. As individuals, we trace our personal ancestry as far back as we can go, and then our lineage disappears. As a species, we trace our complex and divergent ancestry back as far as we can find remains, knowing that what we do not know always greatly exceeds what little we do know. Such are the limits of our knowledge. To be in relation with the Christ who makes us one is simply to live with a consciousness that humanity, so varied in form and often completely cut off from our direct knowledge, is nonetheless connected as an evolutionary whole.

Christ Defines. . .and Undefines

If Christ is the unification and the completion of humanity, then it is also obvious that Christ defines what it means to be human. The theological irony here, however, is that the very definition given to humanity in Christ is really our undefining. Our definition in Christ undercuts our best ideas about ourselves, deconstructing traditional views of human nature. "What does it mean to be human?" is a perpetual question for which there is presently no answer.

For Robert Jenson, what is essential to human nature is our incessant asking what is essential to our nature. He writes that "to be human is constituted, from the creature's side, in asking: What is it to be human?" We are human not because we can answer this question but because we cannot stop asking it. Just to ask the question, however, is an act of self-transcendence precisely because the human questioner refuses to be defined by any previously known answer. Jenson refers to this as our "questioning self-transcendence," a questioning that leads to no answer and thereby frees us from all false answers (Jenson 1997, 64).

For most theologians like Jenson, the reason that human beings cannot answer the question of what it means to be human is because we exist in the image of the unknowable and inexhaustible mystery of God. As Gregory of Nyssa noted centuries ago, to be in the image of a mystery is to be a mystery ourselves. As Jenson puts it, "If there is a *mystery* in created humanity that is the counterpart to the mystery of God, it is such self-transcendence, it is that I am subject of the object I am and object of the subject I am" (Jenson 1997, 64).

Perhaps the most succinct comment on the "undefining" of humanity is found in Karl Rahner's writings. Earlier, we took note of Rahner's cryptic and somewhat playful "definition" of humanity. We are, he says, "an indefinability come to consciousness of itself." Of course we know a great deal about ourselves, whether through introspection or through scientific study. But Rahner insists that "when we have said everything about ourselves that can be described and defined, we have still said nothing about ourselves, unless we have included or implied the fact that we are beings who are referred to the incomprehensible God" (Rahner, 1966a, 107).

Is humanity "an indefinability come to consciousness of itself," as Rahner suggests? Christians will surely hold to various opinions about what it means to say that Christ defines and undefines humanity. However, theologians from Gregory of Nyssa to Karl Rahner who emphasize human indefinability can turn today to the science of human origins, not for direct support, but for a kind of thematic resonance. As we found in our review of recent science, the more we know about our past the more we also recognize that there are no clear lines or boundaries that define humanity. The lines that are drawn are arbitrary, based on the interpretation of findings and not on the data. When we look back in search of ourselves and our origins, we learn of particular forms and sequential relationships. But the more we learn, the more we are met with the puzzling complexities of our past. They confound our attempts to construct any clear narrative. If we look into the deep past, we can find human ancestors but no clear human definitions.

For many people, the same is true when we look deeply into ourselves in the present. Can we truly know other individuals? For that matter, can we know ourselves? The ancient philosophical maxim, "know thyself," seems an impossible task. When we reflect deeply on the conflicting desires and incoherent motives that lie within, we come away agreeing with Augustine. "My heart is incapable of knowing itself," he writes in his commentary on Ps. 39.23. And then, in one of the most famous and distressing lines in his *Confessions*, we read this: "I have become an enigma to myself, and herein lies my sickness" (Augustine, *Confessions* 10.33.50). He cries out to God to save him for his own essential confusion and incoherence, his deep inner self-alienation. The self may see and know and even control all things…except itself.

No matter how much he asks for deliverance, Augustine learns that it must wait but that it is coming. He turns to 1 Cor 13:12, where Paul writes that for now, we see ourselves as if we were looking in a dim mirror. In my final state, Paul says, I will see myself clearly, "face to face," because I will see God, who knows and understands me and in whose loving gaze I know and understand myself. Paul writes: "Now I know in part; then I shall understand fully, even as I have been fully understood." Drawing on this text, Augustine writes: "But when the sight shall have come which is promised anew to us face to face, we shall see this not only incorporeal but also absolutely indivisible and truly unchangeable Trinity far more clearly and certainly than we now see its image which we ourselves are…" (Augustine, *de Trinitate* 15.24). In other words, when we are face to face with God, then the mystery of the Trinity will be more clear than our own persons are to us now.

The theological notion that humans are unknowable is rooted in the belief that we are in the image of the unknowable mystery of God. Theology's claim is more radical than what psychology suggests about how the innermost motives and desires of the self are obscure to us. It goes beyond what the science of human origins suggests when it undercuts all clear definitions about what is "human" and what is not. But if the theological conviction is correct, then what we sense from psychology and what we are now learning from the science of human origins fits appropriately. No matter how we look at ourselves, what we want to know most about ourselves is hidden from view. What does it mean to be human? No matter how we ask the question, theologically or scientifically, the answer eludes us.

Christ of the Future

If what we are now is hidden, how much more hidden is what we shall become. When we look back over millions of years of human evolution, we see the profound changes of the past that make us what we are in the present. We can see how tools played a role in changing our ancestors. They made tools and changed their environments, and this changed them. We are beginning to see how this happened in the past. But when we try to look into the future, about the only thing that seems certain is that new technologies will be developed, probably at an accelerating pace.

How new technologies will change us is completely hidden from view. Our earliest technology altered the local environmental niche. The impact of the technologies of today and tomorrow reach far beyond local effects. Their reach extends to the whole earth and its endangered ecosystem. We have even begun to stretch deep into the solar system and slightly beyond. Even more to the point, our technology now reaches deep within our own bodies, touching neurons and DNA sequences, not just in order to heal through medicine but to modify what we do not like and even to enhance our performance and our abilities. These technologies of human enhancement will continue to be developed and put to use. They will grow in their power and sophistication. Recent advances now point to a day in which the capacities of human beings may be augmented, not just by the new technologies that surround us or are inserted into our bodies and brains, such as pharmaceutical products or digital devices, but by the direct modification of our DNA that we pass to future generations as evolved DNA gives way to edited DNA.

We humans are products of evolution. We are limited biologically by the processes that made us. We could be different. And very likely, because of technology more than anything, we will soon be different. One of the key debates of our time is how far we should use technology to modify or enhance ourselves. What all this means theologically is a question to be explored elsewhere. For now, our point is simple. If we think our past is confusing and largely hidden, just ponder for a moment what lies ahead, when the biological constraints we inherit from the past bend before the technologies we invent in the future.

Despite all that is hidden from our view, we can still say a few things about the humanity of the future. The first is that we must not exaggerate our importance in the scheme of things. As a species, we are still very new on the planet. We ourselves are vulnerable, and yet we are powerful enough to put life on the planet at risk. It took evolution seven million years to make us. If we survive very long, perhaps something like 7,000 years or one tenth of one percent of the time it took to produce us, we are sure to change, not just through biological evolution but through technology.

For the Christian, the importance of our humanity is nothing compared to the glory and majesty of God. When we compare ourselves to the totality of the cosmos as a whole, we humans are small and unimportant. As Jürgen Moltmann puts it, the "human being is not the meaning and purpose of evolution." The vast cosmogenic process has generated human intelligence, but this may already have happened many times in our cosmos, perhaps at levels of wisdom and spiritual awareness that far exceed anything achieved by human beings on planet Earth.

We are right to say with confidence that God loves us. However, we may not say that God creates everything just to create us. The universe, almost incomprehensibly vast, does not exist just so we may evolve. According to Moltmann, "cosmogenesis is not bound to the destiny of human beings. The very reverse is true: the destiny of human beings is bound to the cosmogenesis. Theologically speaking, the meaning and purpose of human beings is to be found in God" (Moltmann 1985, 197). We are a means toward God's ends.

Over recent centuries, theology has fully accepted the idea that our planet is not the fixed center of the cosmos, the point around which everything revolves. The solar system is not geocentric, and the galaxy is not heliocentric. We are still struggling, perhaps, with the anthropocentric correlate of these ideas. We human beings are not the central purpose of the cosmos or even of the planet, the privileged point of consciousness for which everything else serves merely as stage props.

And yet it is also true that at least on this planet, human beings have arisen from the evolutionary process until we stand above it, unique in our conscious awareness and in our powers to alter the natural world. Seen against the long history of our origins, our awakening to consciousness and our sophisticated powers are new, coming only in the past few thousand years. Thinking about this, Karl Rahner suggests that traditional theology was wrong when it assumed that Christ comes late in time, near the end of human history. Ever since the earliest days of the church, Christians saw Christ as the one who inaugurates a new era. But they added quickly that the new era would last a generation or two at most. Our perspective is different. Counting back from today, almost 100 generations have passed since the time of Christ. How many thousands upon thousands of generations of future humans might there be?

Thanks to the science of human origins and the technologies of human enhancement, our thinking about the temporal horizons of our humanity has been stretched both backward and forward. Looking back over millions of years of hominin ancestry, or even just a few thousand years of recorded history, we wonder what lies ahead. What would it mean today for Christians to think that Christ is inaugurating a new era even a fraction as long as our previous hominin evolutionary history? It would suggest that in terms of the fullest possible meaning of humanity, are we not at the end but only scarcely now at the beginning.

According to Rahner, "Today we believe that we know a history of humanity which stretches several hundred times further back into the past than had been imagined in the old days, and we get the impression that, after a very long and up till now almost stagnant starting period, humanity has a history before it, a history whose future in this world has only just begun. Hence, whereas previously one had the impression that God had entered the world through the Incarnation of his Word in the evening of world-history, we now get the impression that (in terms of large periods) he came approximately at the moment when the history of man's active self-possession and of his knowing and active self-steering of history was just beginning" (Rahner 1966b, 189).

Is it right to say that after all our long history, we are only now "just beginning"? We look back seven million years from the time of our divergence from the ancestral line that led to chimpanzees. We note the slow changes that made us only slightly different from our primate cousins. But do the accumulated differences of the past bring us now to a launching point when the most radical changes are only set to begin?

Rahner seems to think so. For him, our present era, what we often call "the common era" that stretches back 2,000 years, is best seen as the staging ground for a new age. He writes that for the Christian, Christ is "the start of other, even intramundane ages of humanity, at the very beginning of this epoch. This means that—beginning with Christ and including also the modern age and the future planetary age, with its higher social organization, an age [is coming] in which man will gain ever greater control over nature and regulate it more and more…" (Rahner 1966b, 190).

To what end? If the true purpose or end of humanity is Jesus Christ, then we recognize that our future form is hidden. We also recognize that our role is to serve the cosmic purposes of the creator. A fascinating text in this regard is found in the Epistle to Ephesians. There we read that God "has made known to us in all wisdom and insight the mystery of his will, according to his purpose which he set forth in Christ as a plan for the fullness of time, to unite all things in him, things in heaven and things on earth" (Eph 1:9-10). What exactly might it mean for all things to be brought to unity in God through Christ?

Rich interpretations of these texts are offered by theologians of the Eastern Church, ranging from Maximus the Confessor (c. 580-662) to the Romanian theologian Dumitru Stăniloae (1903-1993). In a remarkable passage, Stăniloae suggests that human creativity including technology exists to serve what Ephesians describes as God's "plan for the fullness of time." For God to bring the whole creation into unity with the Creator, the creation itself must be transformed or "spiritualized," as Stăniloae puts it. He writes: "The world was created in order that man, with the aid of the supreme spirit, might raise the world up to a supreme spiritualization, and this to the end that human beings might encounter God within a world that had become fully spiritualized through their own union with God. The world is created as a field where, through the world, man's free work can meet God's free work with a view to the ultimate and total encounter that will come about between them."

Stăniloae is clear that God achieves this cosmic transformation with the critical involvement of humanity. It is as if God has made us for this: to transform the cosmos so that with the whole creation, we may come at last to the radiance of oneness with God. Our action is needed, but by itself it is not sufficient. "For if man were the only one freely working within the world," Stăniloae writes, "he could not lead the world to a complete spiritualization, that is, to his own full encounter with God through the world. God makes use of his free working within the world in order to help man, so that through man's free work both he and the world may be raised up to God and so that, in cooperation with man, God may lead the world toward that state wherein it serves as a means of perfect transparency between man and himself" (Stăniloae 2000, 59).

What, then, is the "end" of Adam and Eve, of all forms of humanity that have ever lived and ever will live? It is this:

To the end that the cosmos might achieve its "supreme spiritualization" until it is wholly transformed into "perfect transparency" and united at last with God, and not for ourselves but for the eternal glory of the Creator, we humans have emerged on this earth.

References

Ackermann, R. R. et al. 2015. The hybrid origin of "modern" humans. *Evolutionary Biology*: 1-11.

Alemseged, Z. et al. 2006. A juvenile early hominin skeleton from Dikika, Ethiopia. *Nature* 443.7109: 296-301.

Alves, I. et al. 2012 Genomic data reveal a complex making of humans. *PLoS Genet* 8.7: e1002837.

Antón, S. C., and J. J. Snodgrass. 2012. Origins and evolution of genus Homo. *Current Anthropology* 53.S6: S479-S496.

Antón, S. C. et al. 2014. Evolution of early Homo: An integrated biological perspective. *Science* 345.6192: 1236828.

Argue, D. et al. 2006. Homo floresiensis: Microcephalic, pygmoid, australopithecus, or homo? *Journal of Human Evolution* 51(4): 360-74.

Asfaw, B. 1999. Australopithecus garhi: A new species of early hominid from Ethiopia. *Science* 284: 629-35.

Aubert, M. et al. 2014. Pleistocene cave art from Sulawesi, Indonesia. *Nature* 514 (7521): 223-7.

Aubert, M. et al. 2012. Confirmation of a late middle pleistocene age for the Omo Kibish 1 cranium by direct uranium-series dating. *Journal of Human Evolution* 63 (5): 704-10.

Balter, M. 2010. Candidate human ancestor from South Africa sparks praise and debate. *Science* 328.5975: 154-155.

Bantum, B. 2010. *Redeeming Mulatto: A Theology of Race and Christian Hybridity*. Waco: Baylor University Press.

Bauckham, R. 1995. Tamar's ancestry and Rahab's marriage: Two problems in the Matthean genealogy. *Novum Testamentum* 37 (4): 313-29.

Bedaso, Z. K. et al. 2013. Dietary and paleoenvironmental reconstruction using stable isotopes of herbivore tooth enamel from middle pliocene Dikika, Ethiopia: Implication for Australopithecus afarensis habitat and food resources. *Journal of Human Evolution* 64 (1): 21-38.

Benazzi, S. et al. 2015. The makers of the Protoaurignacian and implications for Neandertal extinction. *Science* 348 (6236): 793-796.

Berger, L. R. 2012. Australopithecus sediba and the earliest origins of the genus Homo. *Journal of Anthropological Sciences* 90: 117-31.

Berger, L. R. 2013a. The mosaic nature of Australopithecus sediba. introduction. *Science* 340 (6129): 163-5.

Berger, L. R. 2013b. Bones of contention: shifting paradigms in human evolution with the skeletons of Australopithecus sediba. Unpublished lecture available at https://www.wits.ac.za/media/news-migration/files/Lee%20BergerInaugural%20Lecturepdf.pdf.

Berger, L. R. et al. 2010. Australopithecus sediba: A new species of homo-like australopith from South Africa. *Science* 328 (5975): 195-204.

Berna, F. et al. 2012. Microstratigraphic evidence of in situ fire in the Acheulean strata of Wonderwerk Cave, Northern Cape province, South Africa. *Proceedings of the National Academy of Sciences* 109 (20): E1215-E1220.

Beyene, Y. et al. 2013. The characteristics and chronology of the earliest Acheulean at Konso, Ethiopia. *Proceedings of the National Academy of Sciences* 110 (5): 1584-1591.

Brown, K. S. et al. 2012. An early and enduring advanced technology originating 71,000 years ago in South Africa. *Nature* 491 (7425): 590-3.

Brunet, M. et al. 2002. A new hominid from the upper Miocene of Chad, Central Africa. *Nature* 418 (6894): 145-151.

References

Campbell, M. C. et al. 2014. The peopling of the African continent and the diaspora into the new world." *Current opinion in genetics & development* 29 (2014): 120-132.

Caron, F. et al. 2011. The reality of Neandertal symbolic behavior at the Grotte du Renne, Arcy-sur-Cure, France. *PloS One* 6 (6): e21545.

Crompton, R.H. et al. 2012. Human-like external function of the foot, and fully upright gait, confirmed in the 3.66 million year old Laetoli hominin footprints by topographic statistics, experimental footprint-formation and computer simulation. *Journal of the Royal Society* 9 (69): 707-19.

Culotta, E. 1999. A new human ancestor? *Science* 284 (5414): 572-573.

Curnoe, D. et al. 2015. A hominin femur with archaic affinities from the late Pleistocene of southwest China. *PloS One* 10 (125): e0143332.

Darwin, C. 1882. *The Descent of Man, and Selection in Relation to Sex* (2nd ed.). London: John Murray.

De Heinzelin, J. et al. 1999. Environment and behavior of 2.5-million-year-old Bouri hominids. *Science* 284 (5414): 625-629.

d'Errico, F. et al. 2009. Additional evidence on the use of personal ornaments in the Middle Paleolithic of North Africa. *Proceedings of the National Academy of Sciences* 106 (38): 16051-16056.

d'Errico, F. et al. 2012. Technological, elemental and colorimetric analysis of an engraved ochre fragment from the Middle Stone Age levels of Klasies River Cave 1, South Africa. *Journal of Archaeological Science* 39 (4): 942-952.

Deschamps, M., et al. 2016. Genomic signatures of selective pressures and introgression from archaic hominins at human innate immunity genes. *The American Journal of Human Genetics* 98 (1): 5-21.

DeSilva, J. M. 2011. A shift toward birthing relatively large infants early in human evolution. *Proceedings of the National Academy of Sciences* 108 (3): 1022-1027.

Diez-Martín, F., et al. 2015. The origin of the Acheulean: the 1.7 million-year-old site of FLK West, Olduvai Gorge (Tanzania). *Scientific Reports* 5.

Douze, K. et al. 2015. Techno-cultural characterization of the MIS 5 (c. 105–90 Ka) lithic industries at Blombos Cave, Southern Cape, South Africa. *PloS One* 10 (11): e0142151.

Falk, D. R. 2004. *Coming to Peace with Science: Bridging the Worlds between Faith and Biology*. InterVarsity Press.

Ferring, R. et al. 2011. Earliest human occupations at Dmanisi (Georgian Caucasus) dated to 1.85–1.78 Ma. *Proceedings of the National Academy of Sciences* 108 (26): 10432-10436.

Fu, Q. et al. 2015. An early modern human from Romania with a recent Neanderthal ancestor. *Nature* 524(7564): 216-219.

Fu, Q. et al. 2016. The genetic history of Ice Age Europe. *Nature* 534 (7606): 200-5.

Gavrilets, S. 2012. Human origins and the transition from promiscuity to pair-bonding. *Proceedings of the National Academy of Sciences* 109 (25): 9923-9928.

Gibbons, A. 2013. Ardi's a Hominin—But How Did She Move? *Science* 340 (6131): 427.

Gibbons, A. 2009. A new kind of ancestor: Ardipithecus unveiled. *Science* 326 (5949): 36-40.

Green, D. J. and Z. Alemseged. 2012. Australopithecus afarensis scapular ontogeny, function, and the role of climbing in human evolution. *Science* 338 (6106): 514-7.

Gregory [of Nyssa]. "On the Making of Man." In *Gregory of Nyssa: The Nicene and PostNicene Fathers*, vol. 5, translated by W. Moore and H. A. Wilson. Grand Rapids: Eerdman's, 1988.

Groucutt, H. S. et al. 2015. Rethinking the dispersal of Homo sapiens out of Africa. *Evolutionary Anthropology: Issues, News, and Reviews* 24 (4): 149-164.

References

Gunz, P. et al. 2010. Brain development after birth differs between Neanderthals and modern humans. *Current Biology* 20 (21): R921-R922.

Haile-Selassie, Y. et al. 2010. An early Australopithecus afarensis postcranium from Woranso-Mille, Ethiopia. *Proceedings of the National Academy of Sciences* 107 (27): 12121-6.

Haile-Selassie, Y. et al. 2010. New hominid fossils from Woranso-Mille (Central Afar, Ethiopia) and taxonomy of early Australopithecus. *American Journal of Physical Anthropology* 141 (3): 406-17.

Haile-Selassie, Y. et al. 2012. A new hominin foot from Ethiopia shows multiple Pliocene bipedal adaptations. *Nature* 483 (7391): 565-569.

Haile-Selassie, Y. et al. 2015. New species from Ethiopia further expands Middle Pliocene hominin diversity. *Nature* 521 (7553): 483-488.

Hammer M. F. et al. 2011. Genetic evidence for archaic admixture in Africa. *Proceedings of the National Academy of Sciences* 108 (37): 15123-8.

Harmand, S. et al. 2015. 3.3-million-year-old stone tools from Lomekwi 3, West Turkana, Kenya. *Nature* 521 (7552): 310-5.

Harrison, T. 2010. Apes among the tangled branches of human origins. *Science* 327 (5965): 532-534.

Hawks, J. and M. H. Wolpoff. 2003. Sixty years of modern human origins in the American Anthropological Association. *American Anthropologist* 89-100.

Hawks, J. et al. 2015. Comment on "Early Homo at 2.8 Ma from Ledi-Geraru, Afar, Ethiopia." *Science* 348 (6241): 1326-1326.

Higham, T. et al. 2012. Testing models for the beginnings of the Aurignacian and the advent of figurative art and music: the radiocarbon chronology of Geißenklösterle. *Journal of Human Evolution* 62 (6): 664-676.

Hublin, J. -J. 2014. Paleoanthropology: Homo erectus and the limits of a paleontological species. *Current Biology* 24 (2): R82-R84.

International Theological Commission. 2004. *Communion and Stewardship: Human Persons Created in the Image of God*. Origins 34.15.

Irenaeus [of Lyon]. *St. Irenaeus of Lyons against the Heresies*. Trans. by D. J. Unger and J. J. Dillon. New York, N.Y: Paulist Press, 1992.

Jaubert, J. et al. 2016. Early Neanderthal constructions deep in Bruniquel Cave in southwestern France. *Nature* 534 (7605): 111-114.

Jenson, R. W. 1997. *Systematic Theology: Vol. 2*. New York: Oxford University Press.

John Paul II [Pope]. 1996. "Truth cannot contradict truth." Address to the Pontifical Academy of Sciences.

Joordens, J. C. A. et al. 2015. Homo erectus at Trinil on Java used shells for tool production and engraving. *Nature* 518 (7538): 228-231.

Kimbel, W. H. et al. 2006. Was Australopithecus anamensis ancestral to A. afarensis? A case of anagenesis in the hominin fossil record. *Journal of Human Evolution* 51 (2): 134-52.

Kimbel, W. H. and L. K. Delezene. 2009. "Lucy" redux: a review of research on Australopithecus afarensis. *American Journal of Physical Anthropology* 140 (Suppl 49): 2-48.

Kinnaman, D. and A. Hawkins. 2011. *You Lost Me: Why Y Christians are Leaving Church... and rethinking faith*. Baker Books.

Kivell, T. L. and D. Schmitt. 2009. Independent evolution of knuckle-walking in African apes shows that humans did not evolve from a knuckle-walking ancestor. *Proceedings of the National Academy of Sciences* 106 (34): 14241-14246.

Kivell, T. L. 2015. Evidence in hand: recent discoveries and the early evolution of human manual manipulation. *Philosophical Transactions of the Royal Society B: Biological Sciences* 370 (1682): 20150105.

Kuhlwilm, M. et al. 2016. Ancient gene flow from early modern humans into Eastern Neanderthals. *Nature* 530 (7591): 429-433.

References

Leakey, M. G. et al. 2012. New fossils from Koobi Fora in northern Kenya confirm taxonomic diversity in early Homo. *Nature* 488 (7410): 201-4.

Lepre, C. J. et al. 2011. An earlier origin for the Acheulian. *Nature* 477 (7362): 82-85.

Liu, W. et al. 2015. The earliest unequivocally modern humans in southern China. *Nature* 526 (7575): 696-9.

Livingstone, D. N. 2008. *Adam's Ancestors: Race, Religion, and the Politics of Human Origins*. JHU Press.

Lordkipanidze, D. et al. 2013. A complete skull from Dmanisi, Georgia, and the evolutionary biology of early Homo. *Science* 342 (6156): 326-331.

Lovejoy, C. O. 2009. Reexamining human origins in light of Ardipithecus ramidus. *Science* 326 (5949): 74-74e8.

Lovejoy, C. O. et al. 2009a. The pelvis and femur of Ardipithecus ramidus: the emergence of upright walking. *Science* 326 (5949): 71-71e6

Lovejoy, C. O. et al. 2009b. Combining prehension and propulsion: The foot of Ardipithecus ramidus. *Science* 326 (5949): 72e1-8.

Lovejoy, C. O, and M. A. McCollum. 2010. Spinopelvic pathways to bipedality: why no hominids ever relied on a bent-hip–bent-knee gait. *Philosophical Transactions of the Royal Society B: Biological Sciences* 365 (1556): 3289-3299.

McPherron, S. P. et al. 2010. Evidence for stone-tool-assisted consumption of animal tissues before 3.39 million years ago at Dikika, Ethiopia. *Nature* 466 (7308): 857-860.

Maricic, T. et al. 2013. A recent evolutionary change affects a regulatory element in the human FOXP2 gene. *Molecular Biology and Evolution* 30 (4): 844-852.

Mendez, F. L. et al. 2016. The divergence of Neandertal and modern human Y chromosomes. *The American Journal of Human Genetics* 98 (4): 728-734.

Meyer, M et al. 2012. A high-coverage genome sequence from an archaic Denisovan individual. *Science* 338 (6104): 222-226.

Meyer, M. et al. 2014. A mitochondrial genome sequence of a hominin from Sima de los Huesos. *Nature* 505 (7483): 403-406.

Meyer, M. et al. 2016. Nuclear DNA sequences from the Middle Pleistocene Sima de los Huesos hominins. *Nature* 531 (7595): 504-507.

Moltmann, J. 1985. *God in Creation: A New Theology of Creation and the Spirit of God.* San Francisco: Harper & Row.

Morimoto, N. et al. 2012. Shared human-chimpanzee pattern of perinatal femoral shaft morphology and its implications for the evolution of hominin locomotor adaptations. *PloS One* 7 (7): e41980.

Moritz, J. M. 2016. *Science and Religion: Beyond Warfare and Toward Understanding.* Anselm Academic.

Muttoni, G. et al. 2010. Human migration into Europe during the late Early Pleistocene climate transition. *Palaeogeography, Palaeoclimatology, Palaeoecology* 296 (1): 79-93.

Nature (editorial). 2016. Humanity's Forgotten Family. *Nature* 534 (7606):151.

Nigst, P. R. et al. 2014. Early modern human settlement of Europe north of the Alps occurred 43,500 years ago in a cold steppe-type environment. *Proceedings of the National Academy of Sciences* 111 (40): 14394-14399.

Pääbo, Svante. 2015. The diverse origins of the human gene pool. Nature Review Genetics 16 (6): 313-314.

Pannenberg, W. 2001. *Systematic Theology: Vol. 2.* Trans. by G. W. Bromiley. Grand Rapids, Mich.: Eerdmans, 2001.

Peters, T. and M. Hewlett. 2003. *Evolution from Creation to New Creation: Conflict, Conversation, and Convergence.* Nashville, Tenn.: Abingdon.

References

Pickering, R. et al. 2011. Australopithecus sediba at 1.977 ma and implications for the origins of the genus Homo. *Science* 333 (6048): 1421-3.

Pico della Mirandola, Giovanni. *Oration on the Dignity of Man*. Chicago: Regnery Gateway, 1956.

Pius XII [Pope]. 1950. *Humani Generis* [Encyclical Letter]. Catholic Truth Society, 1959.

Profico, A. et al. 2016. Filling the gap: human cranial remains from Gombore II (Melka Kunture, Ethiopia; ca. 850 ka) and the origin of Homo heidelbergensis. *Journal of Anthropological Sciences* 94: 1-24.

Quiles, Anita, et al. 2016. A high-precision chronological model for the decorated Upper Paleolithic cave of Chauvet-Pont d'Arc, Ardèche, France. *Proceedings of the National Academy of Sciences* 113(17): 4670-4675.

Rahner, K. 1966a. On the theology of the incarnation. *Theological Investigations*, v. 4. Baltimore: Helicon Press.

Rahner, K. 1966b. Christology within an evolutionary view of the world. *Theological Investigations*, v. 5. London: Darton, Longman & Todd.

Rahner, K. 1981. Unity of the church—unity of mankind. *Theological Investigations*, v. 20. London: Darton, Longman & Todd.

Raichlen, D. A. et al. 2008. The Laetoli footprints and early hominin locomotor kinematics. *Journal of Human Evolution* 54 (1): 112-117.

Reich, D. et al. 2010. Genetic history of an archaic hominin group from Denisova Cave in Siberia. *Nature* 468 (7327): 1053-1060.

Reich, D. et al. 2011. Denisova admixture and the first modern human dispersals into Southeast Asia and Oceania. *The American Journal of Human Genetics* 89 (4): 516-28.

Rendu, W. et al. 2014. Evidence supporting an intentional Neandertal burial at La Chapelle-aux-Saints. *Proceedings of the National Academy of Sciences* 111 (1): 81-86.

Reno, P. L. and C. O. Lovejoy. 2015. From Lucy to Kadanuumuu: balanced analyses of Australopithecus afarensis assemblages confirm only moderate skeletal dimorphism. *PeerJ* 3: e925.

Reyes-Centeno, H. et al. 2015. Testing modern human out-of-Africa dispersal models and implications for modern human origins. *Journal of human evolution* 87: 95-106.

Richmond, B. G. and W. L. Jungers. 2008. Orrorin tugenensis femoral morphology and the evolution of hominin bipedalism. *Science* 319 (5870): 1662-1665.

Robson, S. L. and B. Wood. 2008. Hominin life history: reconstruction and evolution." *Journal of Anatomy* 212 (4): 394-425.

Rodríguez-Vidal, J. et al. 2014. A rock engraving made by Neanderthals in Gibraltar." *Proceedings of the National Academy of Sciences* 111 (37): 13301-13306.

Roebroeks, W. and P. Villa. 2011. On the earliest evidence for habitual use of fire in Europe. *Proceedings of the National Academy of Sciences* 108 (13): 5209-5214.

Roebroeks, W. and M. Soressi. 2016. Neandertals revised. *Proceedings of the National Academy of Sciences* 113 (23): 6372-9.

Sankararaman, S. et al. 2012. The date of interbreeding between Neandertals and modern humans. *PLoS Genetics* 8 (10): e1002947.

Sarchet, P. 2015. New species of early human was Lucy's neighbour in Africa." *New Scientist* (May 27, 2015), available at https://www.newscientist.com/article/dn27604-new-species-of-early-human-was-lucys-neighbour-in-africa/.

Sarmiento, E. E. 2010. Comment on the paleobiology and classification of Ardipithecus ramidus. *Science* 328 (5982): 1105-1105.

Schleiermacher, F. D. E. *The Christian Faith.* Trans. by H. R. MacKintosh and J S. Stewart. New York: Harper, 1963.

Schwartz, G. T. 2012. Growth, development, and life history throughout the evolution of Homo. *Current Anthropology* 53 (S6): S395-S408.

Senut, B. et al. 2001. First hominid from the Miocene (Lukeino formation, Kenya). *Comptes Rendus de l'Académie des Sciences-Series IIA-Earth and Planetary Science* 332 (2): 137-144.

Shea, J. J. 2010. Stone Age visiting cards revisited: a strategic perspective on the lithic technology of early hominin dispersal. *Out of Africa I*. Springer Netherlands. 47-64.

Shea, J. J. 2011. Refuting a myth about human origins. *American Scientist* 99 (2):128.

Simonti, C. N. et al. 2015. The phenotypic legacy of admixture between modern humans and Neandertals. *Science* 351 (6274): 737-741.

Smedt, J. and H. Cruz. 2014. The *imago Dei* as a work in progress: A perspective from paleoanthropology. *Zygon* 49 (1): 135-56.

Snow, D. R. 2013. Sexual dimorphism in European Upper Paleolithic cave art. *American Antiquity* 78 (4): 746-761.

Soressi, M. et al. 2013. Neandertals made the first specialized bone tools in Europe. *Proceedings of the National Academy of Sciences* 110 (35): 14186-14190.

Soressi, M. 2016. Archaeology: Neanderthals built underground. *Nature* 534 (7605): 43-44.

Stăniloae, D. 1994. *The Experience of God*. Vol. 2. *The World: Creation and Deification*. Trans. by I. Ionita and R. Barringer. Brookline, Mass.: Holy Cross Orthodox Press.

Stoneking, M., and R. L. Cann. 1987. Mitochondrial DNA and human evolution. *Nature* 325 (6099): 31-6.

Stout, D. 2011. Stone toolmaking and the evolution of human culture and cognition. *Philosophical Transactions of the Royal Society of London B: Biological Sciences* 366 (1567): 1050-1059.

Stringer, C. 2014. Why we are not all multiregionalists now. *Trends in Ecology & Evolution* 29 (5): 248-251.

Stringer, C. 2015. The many mysteries of Homo naledi. *eLife* 4: e10627.

Sutikna, T. et al. 2016. Revised stratigraphy and chronology for Homo floresiensis at Liang Bua in Indonesia. *Nature* 532 (7599): 366-9.

Suwa, G. et al. 2009. Paleobiological implications of the Ardipithecus ramidus dentition. *Science* 326 (5989): 94-9

Tassi, F. et al. 2015. Early modern human dispersal from Africa: genomic evidence for multiple waves of migration." *Investigative Genetics* 6 (1): 1.

Tattersall, I. 2012. *Masters of the Planet: The Search for our Human Origins*. Macmillan, 2012.

Triska, P. et al. 2015. Extensive admixture and selective pressure across the Sahel Belt. *Genome Biology and Evolution* 7 (12): 3484-3495.

Vallverdú, J. et al. 2014. Age and date for early arrival of the Acheulian in Europe (Barranc de la Boella, la Canonja, Spain). *PloS One* 9 (7): e103634.

van den Bergh, G. D. et al. 2016. Homo floresiensis-like fossils from the early Middle Pleistocene of Flores. *Nature* 534 (2016): 245-248.

Vanhaeren, M. et al. 2006. Middle Paleolithic shell beads in Israel and Algeria. *Science* 312 (5781): 1785-1788.

Van Huyssteen, J. W. 2006. *Alone in the World?: Human Uniqueness in Science and Theology. The Gifford Lectures*. Grand Rapids, Mich.: Eerdmans.

Vernot, B. et al. 2016. Excavating Neandertal and Denisovan DNA from the genomes of Melanesian individuals. *Science* 352 (6282): 235-239.

Villmoare, B. et al. 2015. Early Homo at 2.8 Ma from Ledi-Geraru, Afar, Ethiopia. *Science* 347 (6228): 1352-1355.

References

Walker, M. J., et al. 2016. A View from a Cave: Cueva Negra del Estrecho del Río Quípar (Caravaca de la Cruz, Murcia, Southeastern Spain). Reflections on Fire, Technological Diversity, Environmental Exploitation, and Palaeoanthropological Approaches. *Human Evolution* 31.

Walton, J. H. 2010. *The Lost World of Genesis One: Ancient Cosmology and the Origins Debate*. InterVarsity Press.

Wan, S. -K. 2000. Does diaspora identity imply some sort of universality? An Asian-American reading of Galatians. In *Interpreting Beyond Borders*, ed by F. F. Segovia. Sheffield: Sheffield Academic Press.

Ward, C. V. et al. 2011. Complete fourth metatarsal and arches in the foot of Australopithecus afarensis. *Science* 331 (6018): 750-3.

White, T. D. et al. 2003. Pleistocene Homo sapiens from Middle Awash, Ethiopia. *Nature* 423 (6941): 742-747.

White, T. D. et al. 2006. Asa Issie, Aramis and the origin of Australopithecus. *Nature* 440 (7086) 883-9.

White, T. D. et al. 2009. Ardipithecus ramidus and the paleobiology of early hominids. *Science* 326 (5949): 64-86.

White, T. D. et al. 2015. Neither chimpanzee nor human, Ardipithecus reveals the surprising ancestry of both. *Proceedings of the National Academy of Sciences* 112 (16): 4877-4884.

Wilcox, D. L. 2016. A Proposed Model for the Evolutionary Creation of Human Beings: From the Image of God to the Origin of Sin. *Perspectives on Science & Christian Faith* 68 (1).

Wilkins, J. et al. 2012. Evidence for early hafted hunting technology. *Science* 338 (6109): 942-946.

Wolpoff, M. H. 2009. How Neandertals inform human variation. *American Journal of Physical Anthropology* 139 (1): 91-102.

Wolpoff, M. H. and R. Caspari. 2011. Neandertals and the roots of human recency." *Continuity and discontinuity in the peopling of Europe*. Springer Netherlands. 367-377.

Wood, B. 2011. Did early Homo migrate "out of" or "in to" Africa? *Proceedings of the National Academy of Sciences* 108 (26): 10375-10376.

Wood, B., and T. Harrison. 2011. The evolutionary context of the first hominins. *Nature* 470 (7334): 347-352.

Wood, B. 2014. Fifty years after Homo habilis. *Nature* 508 (7494): 31-33.

Wynn, J. G. et al. 2013. Diet of Australopithecus afarensis from the Pliocene Hadar Formation, Ethiopia. *Proceedings of the National Academy of Sciences* 110 (26): 10495-500.

Made in the USA
San Bernardino, CA
19 June 2018